HANDBOOK OF PRACTICAL ORGANIC MICROANALYSIS

Recommended Methods for Determining Elements and Groups

ELLIS HORWOOD SERIES IN ANALYTICAL CHEMISTRY

EDITORS: Dr. R. A. Chalmers and Dr. Mary Masson, University of Aberdeen

"I recommend that this Series be used as reference material. Its Authors are among the most respected in Europe". *J. Chemical Ed., New York.*

APPLICATION OF ION-SELECTIVE MEMBRANE ELECTRODES IN ORGANIC ANALYSIS
F. BAIULESCU and V. V. COŞOFREŢ, Polytechnic Institute, Bucharest
HANDBOOK OF PRACTICAL ORGANIC MICROANALYSIS
S. BANCE, May and Baker Research Laboratories, Dagenham
ION-SELECTIVE ELECTRODES IN LIFE SCIENCES
D. B. KELL, University College of Wales, Aberystwyth
INORGANIC REACTION CHEMISTRY: SYSTEMATIC CHEMICAL SEPARATION
D. T. BURNS, Queen's University, Belfast, A. G. CATCHPOLE, Kingston Polytechnic, A. TOWNSHEND, University of Birmingham
QUANTITATIVE INORGANIC ANALYSIS
R. BELCHER and A. J. NUTTEN
HANDBOOK OF PROCESS STREAM ANALYSIS
K. J. CLEVETT, Crest Engineering (U.K.) Inc.
AUTOMATIC METHODS IN CHEMICAL ANALYSIS
J. K. FOREMAN and P. B. STOCKWELL, Laboratory of the Government Chemist, London
FUNDAMENTALS OF ELECTROCHEMICAL ANALYSIS
Z. GALUS, Warsaw University
LABORATORY HANDBOOK OF THIN LAYER AND PAPER CHROMATOGRAPHY
J. GASPERIČ, Charles University, Hradec Králové
J. CHURÁČEK, University of Chemical Technology, Pardubice
HANDBOOK OF ANALYTICAL CONTROL OF IRON AND STEEL PRODUCTION
T. S. HARRISON, Group Chemical Laboratories, British Steel Corporation
HANDBOOK OF ORGANIC REAGENTS IN INORGANIC ANALYSIS
Z. HOLZBECHER et al., Institute of Chemical Technology, Prague
ANALYTICAL APPLICATION OF COMPLEX EQUILIBRIA
J. INCZÉDY, University of Chemical Engineering, Veszprém
PARTICLE SIZE ANALYSIS
Z. K. JELÍNEK, Organic Synthesis Research Institute, Pardubice
OPERATIONAL AMPLIFIERS IN CHEMICAL INSTRUMENTATION
R. KALVODA, J. Heyrovský Institute of Physical Chemistry and Electrochemistry, Prague
ATLAS OF METAL-LIGAND EQUILIBRIA IN AQUEOUS SOLUTION
J. KRAGTEN, University of Amsterdam
GRADIENT LIQUID CHROMATOGRAPHY
C. LITEANU and S. GOCAN, University of Cluj
LABORATORY HANDBOOK OF CHROMATOGRAPHIC AND ALLIED METHODS
O. MIKEŠ, Czechoslovak Academy of Sciences, Prague
STATISTICAL THEORY AND METHODOLOGY OF TRACE ANALYSIS
C. LITEANU and I. RÎCĂ, University of Cluj
SPECTROPHOTOMETRIC DETERMINATION OF ELEMENTS
Z. MARCZENKO, Warsaw Technical University
SEPARATION AND ENRICHMENT METHODS OF TRACE ANALYSIS
J. MINCZEWSKI et al., Institute of Nuclear Research, Warsaw
HANDBOOK OF ANALYSIS OF ORGANIC SOLVENTS
V. SEDIVEČ and J. FLEK, Institute of Hygiene and Epidemiology, Prague
FOUNDATIONS OF CHEMICAL ANALYSIS
O. BUDEVSKY, Academy of Medicine, Sofia, Bulgaria
HANDBOOK OF ANALYSIS OF SYNTHETIC POLYMERS AND PLASTICS
J. URBANSKI et al., Warsaw Technical University
ANALYSIS WITH ION-SELECTIVE ELECTRODES
J. VESELÝ and D. WEISS, Geological Survey, Prague
K. ŠTULÍK, Charles University, Prague
ELECTROCHEMICAL STRIPPING ANALYSIS
F. VYDRA, J. Heyrovský Institute of Physical Chemistry and Electrochemistry, Prague
K. ŠTULÍK, Charles University, Prague
B. JULAKOVÁ, The State Institute for Control of Drugs, Prague
ISOELECTRIC FOCUSING METHODS
K. W. WILLIAMS, L. SODERBERG, T. LAAS, Pharmacia Fine Chemicals, Uppsala

HANDBOOK OF PRACTICAL ORGANIC MICROANALYSIS

Recommended Methods for Determining Elements and Groups

S. BANCE, B.Sc., C. Chem., F.R.C.S
Formerly Head of Microanalysis Laboratories
May & Baker Research Institute
Dagenham, Essex

ELLIS HORWOOD LIMITED
Publishers Chichester

Halsted Press: a division of
JOHN WILEY & SONS
New York - Chichester - Brisbane - Toronto

First published in 1980 by

ELLIS HORWOOD LIMITED
Market Cross House, Cooper Street, Chichester, West Sussex, PO19 1EB, England

The publisher's colophon is reproduced from James Gillison's drawing of the ancient Market Cross, Chichester.

Distributors:

Australia, New Zealand, South-east Asia:
Jacaranda-Wiley Ltd., Jacaranda Press,
JOHN WILEY & SONS INC.,
G.P.O. Box 859, Brisbane, Queensland 40001, Australia.

Canada:
JOHN WILEY & SONS CANADA LIMITED
22 Worcester Road, Rexdale, Ontario, Canada.

Europe, Africa:
JOHN WILEY & SONS LIMITED
Baffins Lane, Chichester, West Sussex, England.

North and South America and the rest of the world:
Halsted Press: a division of
JOHN WILEY & SONS
605 Third Avenue, New York, N.Y. 10016, U.S.A.

British Library Cataloguing in Publication Data
Bance, S
Handbook of practical organic micro analysis. –
(Ellis Horwood series in analytical chemistry).
1. Microchemistry – Technique
2. Chemistry, Organic – Technique
I. Title
547'.308'1028 QD272.M5 80–40145
ISBN 0–85312–178–8 (Ellis Horwood Ltd., Publishers)
ISBN 0–470–26972–3 (Halsted Press)

Typeset in Press Roman by Ellis Horwood Ltd.
Printed in Great Britain by Biddles of Guildford.

Table of Contents

CHAPTER 1

Introduction

Since the time, about 1912, that Fritz Pregl introduced micro methods, using 3–5 mg of sample, for the determination of elements such as carbon, hydrogen and nitrogen in organic compounds, the subject (commonly referred to as microanalysis) has seen enormous changes.

The rate of discovery of individual organic substances either by extraction from natural sources or, more often, by synthesis, has increased in an almost explosive fashion.

At first, after Pregl, it was common for the organic chemist to carry out his own microanalysis but it was not long before this work was more often entrusted to a specialist laboratory. Such laboratories found the demand ever increasing and this led to the invention of much more rapid methods and, especially in recent years, to the design of elaborate and costly instruments which will allow a single person to make twenty, thirty or more determinations of carbon, hydrogen and nitrogen per day by almost automatic methods.

Use of such instruments is economically justifiable only where the work load is large enough, and many such laboratories do exist, including some offering contract service, but the present author offers, from his experience of over 30 years of the work, the description of well tried methods (often using very simple procedures and equipment) for all the elements commonly determined.

Carbon, hydrogen and nitrogen are the most frequently determined elements in organic compounds. Consequently C, H & N automatic analysers have proved themselves a great boon to the microanalyst. However, it seems that to make an apparatus which is more rapid it has to be more elaborate and more complicated and consequently more expensive. Not only are they expensive to buy but they cost more to keep in action, for example they usually consume helium and specially pure oxygen all the time. The economics of each situation has to be weighed up. With older methods and no shortage of assistants, carbon, hydrogen and nitrogen can be determined very satisfactorily. The difficulty is often that of obtaining enough assistants possessing sufficient skill and aptitude. In the author's opinion and as a rough guide, as soon as a laboratory finds it has to deal with an average of 10 samples every working day, each requiring C, H & N determination, then the purchase of a C, H & N analyser is advisable even though it may seem under-utilized since it is easily capable of dealing with more than 20 samples a day.

However, as soon as the work load and number of assistants in the laboratory imply that older methods would be too slow to deal with the demand, then consideration must also be given to providing a second automatic analyser as a reserve should the first temporarily break down. This extra expense must be considered. Where the number of samples is regularly an average of more than 20 a day it is probably justified. Laboratories which use these instruments continuously find they have to have someone (or even several people) with an understanding of them readily available, who can detect and remedy faults in the event of break-down. The instruments cannot be expected, like a washing machine or a television set, just to be switched on and to work without trouble.

In any laboratory there may be one of two possible situations. In some laboratories certain determinations, that of phosphorus, for example, will be needed only occasionally, whereas in others the determination may be carried out all day every day. When the latter is the situation, whatever the determination in question, the laboratory then becomes expert and experienced in the method, and its limitations, advantages and the precision to be expected are well known. On the other hand most laboratories need methods, not only for phosphorus but probably also for metals, water, fluorine, boron and silicon, which, although performed only occasionally, will still give good results without much trouble. The following pages describe many such methods. It is necessary to include the fullest details of the procedure in each case so that the occasional determination can be expected to be successful.

Samples submitted for quantitative analysis may have been extracted from natural sources, or more frequently, have been prepared by synthesis. They will either be solid or liquid at laboratory temperature (gaseous products can also be analysed but are not considered here). From the analyst's point of view, however, there are four classes of sample. There are solids and liquids which can be weighed in open vessels in air without change in weight and there are those which change in weight. These may pick up water, oxygen or carbon dioxide from the air or they may lose weight because they evaporate (or sometimes, unintentionally, they contain a solvent which evaporates continuously). Nowadays, usually in a laboratory where there is a continuous demand for repeated determinations of the same kind, routine work is entrusted to young and only partly experienced assistants. They are asked to follow closely defined directions. Such directions therefore must also be of four kinds, not just for the easiest, that is the solid having a weight that is stable in air, but for solids and liquids, stable and unstable.

A quantitative analysis is usually performed in order to identify the sample material; in the case of a natural product it is the first step in assigning to it a possible molecular formula and in the case of a synthetic product it is usually simply confirmation of a postulated formula that is hoped for. To obtain the theoretical figures does not, of course, show which of many possible configurations it may be, but not to obtain theoretical figures does show that the sample is not what it was postulated to be.

In addition, quantitative analysis by an experienced laboratory is often used, although this is not the primary purpose, to show how pure, within limits, a given sample of a known substance is. Although the errors are comparatively high (1% relative for most elements but usually less than this for carbon), an analysis that is approximately the theoretical is usually taken as confirming the identity of the substance. It should be noted that given a C, H & N analysis of good precision, a quite approximate halogen or sulphur figure for example will indicate which whole number of these atoms may be present.

At the present time samples can be examined with a variety of physical instruments. This fact sometimes makes the need for quantitative analysis less but usually this (microanalysis) is relatively so easy to do that the information it provides remains well having.

Infrared spectroscopy (IR for short) helps to confirm or detect the presence of certain groups in the molecule. A very useful rapid routine IR service has been provided for many years in the author's laboratory and found extremely useful. Sometimes it tells the synthetic chemist how far a reaction has proceeded and may thus postpone the need for microanalysis.

Nuclear magnetic resonance (NMR) is a technique, especially as applied to hydrogen atoms in organic substances, which is now in very widespread use and is almost an indispensable part of the organic chemists' armoury. It gives information about the kind of carbon and hydrogen groups present in the sample and can be made to give quantitative results. In spite of the existence of this technique, the description of simple *O*-alkyl, *N*-alkyl and acetyl group and also active hydrogen determinations are included in this book because they are relatively easy to carry out and are sometimes needed. In addition they afford practical demonstrations of certain principles of organic chemistry.

The mass spectrometer (MS) can also provide information about a sample. Mass spectrometers vary in resolving power (and cost) but the highest resolution instruments are not indispensable for routine work. The instrument will often provide an indication of the molecular weight of a sample, rendering the older micro methods of molecular weight determination unnecessary.

Gas liquid chromatography (GLC) is an extremely valuable technique, showing among other things whether a sample is pure (i.e. one component only). It is also used to prepare pure samples, usually liquids at room temperature, but should not be regarded as infallible in this.

Thin-layer chromatography (TLC) is also an almost essential modern technique which can be applied to solid compounds in solution.

Solution chromatography has seen a revival recently in the so-called high-performance liquid chromatography (HPLC) where the process is speeded up by using high pressures to drive a solution of the sample through a column of solid adsorbent. It also can be used either analytically or preparatively.

Thermal analysis in the form of differential thermal analysis (DTA) or differential scanning calorimetry (DSC) can also give valuable information about a sample. Of course, the traditional micro process of observing a sample under the microscope as its temperature is raised or lowered by a heated microscope stage, as in the work of Kofler and Kofler [1], is a form of thermal analysis.

The information obtained by means of all these physical instruments is often extremely valuable but rarely makes unnecessary a quantitative analysis (microanalysis) by the well established methods, which are relatively easy to perform, as described in the pages which follow.

The determination of the amounts of various elements present in organic compounds, the subject of this handbook, is sometimes confused with the other type of microanalysis which can better be called trace analysis. In an absolutely pure organic substance the amount of an element present is substantial, for instance one sulphur atom (a.w. 32) present in a molecule of m.w. of even 1000 would correspond to 3.2% sulphur. Trace analysis on the other hand might be concerned with parts per million concentration levels (1 ppm is 0.0001%). However, it sometimes happens that the microanalyst is asked to determine the percentage of an unintended impurity and although the techniques are not designed for this purpose (the synthetic chemist often does not seem to realize this) the results obtained can sometimes be useful. For instance, chlorine determination by the Carius method (p. 101) can distinguish between 0.2 and 0.3% chlorine or as little as about 0.1% of ash can be measured with reasonable precision (p. 121).

The methods of quantitative organic analysis have developed and improved in rapidity and precision over the years but the most used technique has throughout been that of oxidation of the sample by one means or another without loss of the element to be determined, but with its liberation from organic combination. In other words, the organic compound must be converted into an inorganic species as quickly and easily as possible and then the element of interest can be determined by any method of inorganic chemistry.

Three main techniques using this principle are exemplified in this book.

(1) The first is the complete combustion of the sample by heating it in oxygen in such a way that all the element of interest is converted into a gaseous compound and can be carried in the oxygen stream to an absorber where it is collected. Chapter 4 describes this principle for the determination of carbon and hydrogen by the Belcher and Ingram method.

(2) The second well established method is the combustion of the sample in a closed flask full of oxygen, in which all the products are inorganic in form and retained in the flask. This technique, now so widely used, is described in Section 3.2.

(3) The third method, not now very much used, but simple and effective and therefore extremely useful in some special cases, is the micro-Carius method

(heating the sample in a sealed glass tube with nearly anhydrous nitric acid) and this well tried technique is described in Section 3.1. (The method of destruction of the organic substance by heating in a closed metal bomb with sodium peroxide is retained for only the determination of silicon).

The author has never rigidly adhered to strictly micro quantities of sample (3-5 mg) for determinations and sees no reason why maximum precision should not be striven for by taking a suitable sample size, provided that sufficient sample is available. For example, in the case mentioned above of a sample containing 3.2% S, 10 mg would give a titration of 1 ml of 0.01*M* barium chloride, but 20 mg would be preferable, requiring 2 ml. An even larger sample would be desirable with an ordinary semimicroburette and 0.01*M* barium chloride (which is the traditional 'micro' strength.)

The tendency towards use of semimicro sample sizes makes it generally easier for the young and semiskilled to obtain satisfactory results and removes some of the mystique, that is, the 'this can be done only by an expert' attitude to microanalysis.

During and after the second world war, when most supplies from the continent of Europe were stopped, it was difficult to obtain apparatus for microanalysis. In the hope of improving this situation the British Standards Institution was asked to prepare standards for such apparatus and a whole series was eventually published as parts of BS1428. Such things as weighing vessels, crucibles and filters were standardized, but one of the main goals was to specify sizes so that parts would fit together, for example, in combustion trains. The result is that there exist many specifications, some of which are referred to in this book. Even if an individual does not want the article as specified, the B.S. often provides a convenient starting point for describing the article in question (e.g. the author recommends absorption tubes complying in all respects to the BS except for a slightly larger diameter, p. 40).

It was not until about 1946 that quartz combustion tubing became available. Previously combustion glass, a glass with as high a melting point as was available, was used. Heating was usually by coal gas burners, which in one way was probably just as well because temperatures above 750°C inside the tube were difficult to obtain with gas flames and in any case tubes began to melt slightly and to distort at this temperature. Therefore when fused quartz tubing became available electric heaters were introduced, higher temperatures were possible and consequently combustion was often more rapid and efficient.

BS1428 included a specification of what can be regarded as a standard 5 ml or 10 ml semimicro burette with automatic zero, reservoir and pressure-filling device (BS1428: Part D1: 1965).

In many laboratories these have been replaced by piston burettes (where the rotation of a spiral determines the movement of the piston and thus the volume delivered). The author recommends caution with these, as after a time wear may effect their reliability, and has himself reverted to the more traditional burette (which can now be obtained in an excellent demountable form, with jet, tap, graduated portion, etc. all joined by ground-glass joints). In any case a typical 5 ml titration must have a better precision than 0.05 ml (1%), represented roughly by one drop of solution.

An experienced microanalyst always keeps certain generalizations in mind. For example, a combustion train needs to reach a kind of equilibrium and therefore is more reliable when in constant rather than intermittent use, and articles such as filter crucibles, which are washed, heated, cooled and weighed are more likely to have reached constant weight if they have been in use a long time. When a new method is tried out and found satisfactory then much time has probably been spent to reach this stage and any suggestion of changing to yet another method is resisted. This produces a kind of conservatism in the microanalyst. The following chapters offer descriptions of methods that have stood the test of time.

If a certain method has to be corrected for a blank which itself has to be determined experimentally, then if it is contrived that the final measurement, often a titration volume, is always of the same magnitude, the absolute value of the blank will be less important, the calculation factor being the most important feature, and this is nearly always determined by carrying out the determination on known pure substances [2] and thus is empirical. If a determination on an unknown substance results in an unexpectedly large or small titration volume, for example, it is better to repeat it with an amount of sample that will give a more suitable titration volume.

It has been found that assistants are very happy to have available in the laboratory the slow but reliable Carius methods for the confirmation of unexplained or unexpected results. Also, even when automatic C, H & N analysers are in constant use it has been found very helpful, almost a necessity, to retain the Belcher and Ingram carbon and hydrogen apparatus for occasional use (p. 35). Among other advantages it will allow the weighing of an ash for example, and will also give the true hydrogen content of a hydrate which might lose water during the sweep step of an automatic analyser.

All methods used may be checked against pure known organic substances. In fact it is true to say that it is better to calibrate all methods against these substances rather than to use theoretical factors. Substances used for this purpose must be easily obtained pure, remain stable indefinitely in air and be non-hygroscopic. They must also be available for all the common elements determined. Such a list was compiled by a committee of the then Society for Analytical Chemistry, was published in *The Analyst* and is available as a reprint [2]. These substances can be obtained commercially in a suitable state of purity.

The following chapters offer selected methods for the determination of all the common elements. Many others may have been tried over the years but those described here have stood the test of time not only for reliability but also for rapidity and ease of performance.

Any laboratory set up to provide a service of quantitative organic analysis should, right from the start, institute a system for recording and reporting results. Various kinds are used but the main essential is that the one chosen should be systematic, and then conventions can be used which minimize book-keeping and allow analysts to devote the maximum of their time to analysis.

In the author's laboratory the system for many years has included a printed card (3 x 5 in.) which constitutes the request for analysis and accompanies the sample, which is labelled by a unique combination of letters and numbers which refers to that particular sample only. This reference number is, of course, also entered on the card. A different sample, even if thought to be of the same substance, must have a different reference code and number, although this may mean only the addition of a, b, c, ... or 1, 2, 3 . . . . Each card is given a serial number, conveniently printed on it in red by using an enumerator (available commercially).

Results are recorded on the reverse of the card and the card is filed after the analysis is completed and is kept always available for reference in the micro-analysis laboratory.

Results are transmitted to the submitting chemist by slips torn from printed pads. Here again entries are kept to the minimum to save analysts' time. On the slip the chemist is given the analysis number by which the card can be found in the file if required.

A computer system of sample-logging and data-handling for use in a very large industrial control laboratory has been described by Merz [3].

The majority of the chapters which follow are arranged with the following headings and in this order: Summary of the method, Discussion, Apparatus, Reagents, and Procedure.

Reagents. Reagent chemicals are of laboratory-reagent grade throughout unless otherwise stated. The abbreviation A.R. refers to a reagent of analytical grade (e.g. AnalaR, Pronalys), and M.A.R. refers to a microanalytical reagent that conforms to the standards recommended in [2]. The main suppliers of laboratory reagents in the U.K. are BDH Chemicals Ltd., Hopkin and Williams, and May and Baker Ltd.; all three should be able to supply most of the reagents mentioned in this book. They also supply various standardized volumetric solutions (e.g. 'Volusol', 'AVS') and concentrated volumetric solutions (e.g. 'Volucon', 'CVS'. 'Convol'), which are often convenient time-savers in the laboratory, though many analysts will still want to standardize them.

Appendix A gives a list of text books, to which the author makes the fullest acknowledgement.

REFERENCES

[1] L. Kofler and A. Kofler, *Thermo-Mikro Methoden zur Kennzeichnung Organische Stoffe und Stoffegemische,* 3rd Ed., Verlag Chemie, Weinheim Bergstrasse, 1954.

[2] Reference substances and reagents for use in organic micro-analysis, *Analyst*, 1972, **97**, 740.
[3] W. Merz, *Talanta*, 1974, **21**, 481.

CHAPTER 2

Balances and weighing

This handbook is about methods for finding out how much, that is, what percentage, of a certain element (or group of elements) is present in a given sample of an organic substance. Clearly the precision with which the weight of the portion taken for analysis is measured is of the utmost importance. Sometimes less than a milligram is taken, but, at the other extreme a sample weight of as much as 40 mg or even more may be recommended.

Roughly speaking, to achieve results which are not in error by more than one part in a hundred, samples taken for analysis must be weighed with a precision which corresponds to an error no greater than one part in a thousand.

This chapter describes certain balances and weighing techniques which have been found to be satisfactory.

2.1 GENERAL DISCUSSION

One important factor in the foundation of organic microanalysis by Pregl was the Kuhlmann balance, which could weigh as little as 3 mg with an error no greater than 3 μg (the one part in a thousand mentioned above), but at the same time would keep this sensitivity when loaded with about 10 g in each pan, e.g. when weighing absorption tubes as in the determination of carbon and hydrogen. This balance was itself a modification of the balance produced by Bunge in 1880, which was the first capable of weighing with high sensitivity [1].

Then and for some time after, the rest-point of the pointer had to be determined by calculation from the turning points of the pointer attached to the balance beam. Later aperiodic (fully damped) balances appeared and the rest-point of the pointer could be read directly. These balances all had knife edges of agate or other hard material bearing on planes of similar material and became the most important tools of the microanalytical laboratory, and still are, because the object being weighed may be contained in a vessel such as a crucible or absorption tube which may weigh up to 20 g.

In recent years many electrical balances have become available on which quantities of sample of a few milligrams or less can be weighed with adequate precision but these balances are rarely capable of taking loads such as absorption-tubes. They serve a very important purpose, however, because in some modern methods carbon dioxide and water, for instance, no longer have to be weighed in absorption tubes. In addition it is now usual to select a balance which is suitable for use with an automatic elemental analyser and then it often remains devoted solely to that purpose and may be in use all day every day. Such balances, especially those with digital read-out or direct read-in to a computer, though expensive, may become justifiable because they are used repetitively and minimize the risk of mistakes. They then become regarded as part of the automatic analyser.

It may be mentioned that because many laboratories are investing in automatic analysers and using the most up-to-date electronic time-saving balances with them, some of the more traditional but none the less thoroughly reliable microchemical balances are becoming available on the secondhand market. It is interesting to note that old-fashioned free-swinging balances sometimes have a lower standard deviation than modern balances.

When installing an automatic analyser the user will naturally be guided by the manufacturer regarding the choice of a suitable balance, which, as mentioned above, will probably be given no other use than in conjunction with that particular analyser. However, the critical and impartial Electrobalance Survey made by the Elemental Analyser User Forum in London may be mentioned [2], which has shown that some modern balances may give inadequate performance, but this may be attributed to a faulty balance or faulty operation, or both.

It should be noted, however, that various balances have since become available for use with automatic analysers, such as the Mettler M30 and the Perkin–Elmer AD2X.

Most laboratories prefer to entrust the cleaning and maintenance of balances to an expert provided by the manufacturer (usually under contract) but if the user wishes to do it himself he will find the guidance published by Wilson [3] very helpful.

When a sample is weighed, in nearly all cases a vessel has to be weighed with and without the sample (a rare exception might be a button of gold, or some other metal, which could be placed on the balance pan). Consequently if it is desired to check the precision of the weighing it should be done in a way which corresponds to actual practice, for example, by first weighing an empty boat, removing it from the balance, adding a small piece of platinum to represent the sample, replacing the boat on the balance pan, closing the balance and reweighing. The extent to which the weight of the piece of platinum is reproducible then represents the reproducibility of the weighing of a solid sample which is stable in air.

Only by demonstrating the reproducibility (precision) in this way can the operator be really sure of it. A worn or dirty balance, of course, is less likely to be as satisfactory as a clean or new one.

2.2 BALANCES

This section describes a few selected balances as typical examples of their type. Personal preference and laboratory budget will dictate the models to be purchased.

Oertling balance, model 147

This is a microchemical balance which is fully damped, with a pointer scale projected onto a screen at eye level.

It has two rows of ring riders, added or removed by a knob and dial system with the balance door closed; there are weights of 1, 2, 3 and 5 mg and 10, 20, 30 and 50 mg so that any weight up to 99 mg may be added.

The pointer scale is divided into 500 divisions and the sensitivity is adjusted so that the smallest rider, 1 mg, produces full-scale deflection, so that one division is equivalent to 2 μg. With this sensitivity the balance requires great care in use and therefore the model 141/147 is more to be recommended for routine repetitive use (especially when larger sample weights are used as generally recommended in this book).

Oertling balance, model 141/147

This is also an aperiodic microchemical balance. It is useful to have several of them in the balance-room so that each can be devoted to a particular determination.

It is similar to the model 147 but the sensitivity is adjusted so that the pointer scale of 100 divisions corresponds to the deflection caused by the smallest ring rider weight of 1 mg. Hence one division is equivalent to 10 μg and fractions of a division estimated.

The precision of this balance has been shown to be usually better than 0.01 mg with careful use. The balance is aperiodic, that is to say that when the beam is released, it does not swing from side to side, but, because of the damping caused by pistons moving in pots, one above each pan, the beam moves until it reaches the position of equilibrium and then stops. The position of equilibrium is indicated on the scale carried by the pointer, which is projected optically on to a screen at eye level. It is the pointer scale which has 100 divisions each corresponding to 0.01 mg. The maximum load permitted on each pan is 30 g.

The balance pan wires are fitted with a pair of hooks so that an absorption-tube can be placed on the balance in a stable manner (see p. 47). The counterpoise is placed on the hooks of the other pan in a similar manner. These hooks are also used to support weighing sticks (Fig. 2.1, p. 24), when these are used to weigh solids.

Oertling balance, model 146

In this model the pointer scale has 500 divisions which correspond to the deflection caused by a 10 mg ring rider. Therefore each division represents 0.02 mg and the balance is a semimicro model.

It is very well suited to those operations where a larger than normal sample weight is taken, from 10 to 100 mg, and can be used, for example, for Karl Fischer titrations of water and for halogen determinations by the Carius method.

There are four ring riders that can be added automatically, namely, 10, 20, 30 and 50 mg.

The maximum load on each pan is 30 g.

Cahn gram electrobalance (model G)

This balance is suitable for use with the Perkin-Elmer Model 240 C, H & N analyser.

For this purpose a platinum boat of weight about 90 mg is counterpoised by means of a similar boat on the opposite pan and a sample of from 1 to 3 mg is weighed in the boat.

For such amounts of sample the range knob of the balance is set to the 5 mg position and the readings recorded are then twice the actual weight of the sample, e.g. 4246 μg corresponds to a 2.123 mg sample.

The precision is better than 0.001 mg on this range. (Other ranges available are 10, 20, 40, 100 mg.)

To use the balance the empty boat is counterpoised by bringing the pointer of the beam to the cross-wire by turning a knob, then the sample is added and the beam is brought back to the crosswire by turning another knob, which in fact adds an electric current to the electric motor, a third knob being used to measure this additional current, which is proportional to the weight taken.

In a similar way the balance is calibrated against a 5 mg weight. This procedure is made shorter by the use of a calibrate knob.

Cahn electrobalance model M-10

This balance, which is simpler than the Cahn Gram balance, is suitable for the determination of nitrogen on the Coleman Nitrogen Analyser (p. 53).

Ordinary balances

An ordinary analytical balance should be readily available for weighing out solids for standard solutions, solutions of indicators etc. If it is of a modern rapid type valuable time can be saved. A top loading balance is useful for rough work, such as preparation of bench reagents.

2.3 WEIGHING

The knife-edge type balances are best set up in a balance room. Linoleum covers have been found practical for the balance tables, which may consist of heavy concrete tops supported on brick piers which bear on the main structure of the building, although a well constructed wooden structure is equally satisfactory.

Static charges sometimes give trouble, especially in cold weather or with low water content of laboratory air or if nylon or terylene clothing is worn. A humidifier may be turned on in the balance room when thought advisable. An anti-static pistol (designed for use with gramophone records, and based on a piezo-crystal) has recently been found to be useful [4].

Counterpoises

Many soda-glass objects which are repeatedly weighed are provided with counterpoises made of the same glass. Platinum boats are often counterpoised with a twist of nichrome wire (non-magnetic). Counterpoises must match in density the object being weighed, or buoyancy errors can become appreciable if large pressure changes occur during a working day.

Sets of weights

Each microchemical (knife-edge) balance is provided with a box of weights from 1 to 1000 mg, specially adjusted for microanalysis, obtainable from various manufacturers. In the balance room there should also be available a box of good quality analytical weights up to 20 g. A similar box of weights (preferably with a calibration certificate) should be carefully stored so that they are used only for reference purposes. The working weights must be calibrated against each other and all the sets used should be referred to the same weight (at least 10 g) from the standard set.

Weighing vessels

The choice of the type of vessel used in the weighing of a sample for microanalysis depends very much on how the analysis is to be done. Thus, a sample for C and H analysis is weighed in the platinum boat in which it is to be combusted, after the boat has been weighed empty. Similarly, for nitrogen analysis with the Coleman Analyser, the aluminium boat is first weighed empty and then again with the sample. The weighing tubes for Carius analyses (p. 28) and the cigarette-paper and glass cups for Kjeldahl digestions are used in a similar way.

In other analyses, the sample is weighed in a vessel which is used only in the weighing process: then, the vessel is weighed with the sample in it, the sample is transferred to another vessel (or, for oxygen-flask combustions, onto a filter-paper flag) with care to avoid losses, then the vessel is reweighed so that the weight of sample transferred can be found.

A very convenient vessel for weighing samples in this way is a small platinum boat, similar to that described in BS 1428 [5], but about 6 mm shorter in overall length, and weighing about 0.35 g. This is readily cleaned between samples by ignition in a flame, and since cooling is rapid (~ 1 min on a cooling block) it is no inconvenience if only a single boat is available. A simpler alternative is a 'weighing shovel' made from platinum foil, thickness 0.05 mm (copper or other metal foil may sometimes be used), cut to about 10 × 25 mm and with sides folded up along its length.

When the sample has to be transferred to the bottom of a tube, it is necessary to use a vessel with a long handle for weighing. The traditional microanalytical device is the weighing stick [6], shown in Fig. 2.1.

Fig. 2.1 – Weighing stick

The stick, containing the sample, is first placed on the hooks of the balance, with a similar one or a glass rod on the hooks of the other pan. The sample is then tapped out into the analysis vessel and the stick reweighed. The weighing stick is not touched in use with the bare fingers, but only with chamois-leather gloves or chamois-tipped tongs.

Sticks are made of soda and *not* borosilicate glass, to minimize the risk of static charge. For the same reason, sticks should not be dried in the oven, but should be rinsed with acetone after use and left to stand for some time beside the balance before re-use. For this reason, it is useful to have quite a number, say ten, all carefully adjusted to be close to each other in weight so that a closely counterpoised pair can easily be found.

Filling and cleaning the sticks is not easy, however, and a simpler method is to weigh the sample in a platinum boat, then lift the boat with a long handle consisting of a small crocodile clip attached to a rod [7]. The boat can then be emptied into the bottom of the tube, the handle removed, and the boat reweighed.

Weighing of volatile liquids

Volatile liquids can usually be weighed in a small tube drawn out to a capillary which needs to be sealed only if the liquid is very volatile. Standard melting-point tubes made of Pyrex glass are a convenient source of capillaries, which can be made in a small gas flame. They are sealed at one end and drawn out to a not too fine capillary at the other. Their size depends upon how much sample is required. The whole operation has to be mastered by the individual analyst, who usually has some difficulty at first.

Hygroscopic samples

It frequently happens that it is not known that a sample is hygroscopic until it is seen to be gaining weight when in an open vessel, perhaps a platinum boat, on the balance pan (the sample may instead be gaining carbon dioxide or oxygen, but water from the air is much more likely).

If the sample is handled rapidly, approximate results will be obtained and may be of some value. For good results, the samples should be equilibrated in air and re-analysed with possibly an additional determination of water content.

It is often necessary to try to characterize the dried sample. It must then be handled rapidly, possibly in a dry box, and weighed in a closed vessel (e.g. a weighing pig [5]) closed from the air. Similar treatment should be used for samples which are readily oxidized when exposed to the air (in this case a dry box containing an inert atmosphere would be necessary).

Hygroscopic samples occur quite fequently, and they remain a continual source of trouble to the analyst.

Use of the balance

Because of its high sensitivity, the microbalance demands special care in its use if mistakes are to be avoided. Effects that are too small to be detected with an ordinary balance weighing to 0.1 mg may often be detectable with the microbalance and steps must be taken to prevent their occurrence.

The electrical balances are not affected by environmental factors such as vibration, temperature variation etc., but are of restricted use because of the weight-loading limitation. Most other balances are now of the single-pan type, operating at maximum load, and are aperiodic. The two-pan types are also generally aperiodic, and ring-riders on a carrier bar are usually used in preference to the rider that has to be moved along the beam. One of the penalties of the introduction of instrumentation, especially of the automatic type, has been the development of a belief that an automatic instrument cannot make a mistake and that the readings obtained must be true. Unfortunately, such a belief may be mistaken. The older analysts were taught to calibrate and check regularly all the basic tools of their trade, such as sets of weights, balance sensitivities, pipettes and burettes. By doing so they avoided errors that arise from gradual changes in performance or sudden effects arising from accidental but unreported damage. Such care is still needed if completely reliable results are to be obtained. It is therefore necessary to calibrate weights and so on, and to check the balance sensitivity as a matter of routine. The zero-points of single-pan balances are sensitive to temperature change and should be checked before each weighing – an operation that was second nature to users of free-swinging two-pan balances. Differences in temperature between the balance and the object being weighed can cause drift in readings because of convection currents and other effects, and this drift is easier to detect on a free-swinging balance because it appears as unequal decrements in the right and left-hand turning points. With an aperiodic balance it may pass unnoticed, being mistaken for the natural period of the balance. This temperature effect is most marked in the weighing of the absorption tubes in a C and H analysis (p. 47). When a two-pan balance is used, the effect demands that if the operator puts a hand in at one side of the balance, the other hand should be put in the other side for a similar period as a compensating measure. Cross-effects of body-radiation are sometimes observed when two operators are using adjacent balances for any great length of time.

Adsorbed water films on containers may also cause error. All glassware etc. will pick up water vapour from the atmosphere, and unless the amount remains reproducible within the limits detectable by the balance there will be a weighing error if the same object is weighed at intervals. It is this that is partly responsible for the use of counterpoises matching the object in material, size, treatment, etc., the idea being to have a compensating change in the film picked up by the counterpoise. This is especially important when weighing objects such as sintered-disc filters, which have a very large surface area, especially if there are changes in humidity between weighings. The second objective in the use of matching counterpoises is to avoid errors arising from buoyancy effects if the density of the object to be weighed and the weights used is large and atmospheric pressure changes suddenly by a large amount. Changes of up to 50 mmHg or more in a day have been recorded, and can cause errors of up to 25 μg if lead shot is being used in counterpoising. The necessity for a reproducible water vapour film is also the reason for wiping glassware with chamois leather (gently, to avoid causing an electrostatic charge) and leaving it to equilibrate with the atmosphere before weighing. Lack of equilibration will often show up as drift in the balance readings.

Errors in weighing were invariably discussed in the older books on organic microanalysis and elsewhere, and these sources should be consulted for fuller information [8,9].

REFERENCES

[1] H. Roth, *Pregl-Roth Quantitative Organische Mikroanalyse*, 5th Ed., Springer, Vienna, 1947, p. 1.

[2] P. R. W. Baker, D. A. Pantony and B. T. Saunderson, *Proc. Anal. Div. Chem. Soc.*, 1978, **15**, 136.

[3] D. W. Wilson, *Metallurgia*, 1944, **31**, 101; 1945, **32**, 85; 1946, **34**, 219, 279.

[4] M. R. Masson, Personal communication.

[5] B. S. 1428: Part I1: 1963.

[6] B. S. 1428: Part H1: 1960.

[7] M. R. Masson, *Ph.D. Thesis*, University of Aberdeen, 1974.

[8] G. F. Hodsman, *Proc. Intern. Symp. Microchemistry, Birmingham, 1958*, Pergamon, Oxford, 1959, p. 59.

[9] M. Corner, *Proc. Intern. Symp. Microchemistry, Birmingham, 1958*, Pergamon, Oxford, 1959, p. 64.

CHAPTER 3

The decomposition of organic substances

3.1 THE MICRO CARIUS METHOD [1]

The micro or semimicro Carius [2] method can be so valuable in certain cases because of its reliability and simplicity that it is described in full here. It is recommended as providing a very straightforward method for the determination of phosphorus and various metals and occasionally its use is advisable for halogens and sulphur.

The sample, which can weigh from 1 to as much as 45 mg, is decomposed by heating for a few hours at 300°C with 0.3 ml of fuming nitric acid in a sealed glass tube, with or without addition of 50 mg of sodium chloride. After cooling, the tube is opened and the contents are dissolved in water. Carbon and hydrogen will have been converted into carbon dioxide and water and the element to be determined will be present in the fully oxidized form, such as sulphuric acid or phosphoric acid.

3.1.1 Apparatus

Micro Carius tubes. These are made of borosilicate glass, with one closed, hemispherical end, well annealed, having an outside diameter of 9.7-10.5 mm, wall thickness 1.2-1.6 mm and length 350 mm

Carius tube heater. A heater is required which is capable of heating as many as 9 sealed Carius tubes to 280–300°C. This can be made from 9 pieces of brass tubing of 11.5 mm bore and 250 mm in length, brazed together in a block. A strip electric heater such as an electric iron heater can be clamped to the underside. The whole can be wrapped in glass wool and enclosed in sheet aluminium casing so that the tubes are sloping slightly with the mouths uppermost. A hinged lid over the mouths of the tubes, packed with glass wool lagging, prevents too rigid enclosure in the event of an explosion but also keeps the heat in. Suitable electric connections are made and a variable transformer is used to provide the correct temperature. The setting of the transformer is found by trial and error. It usually takes about 30 min to reach 280°C. Such a home-made heater has been found satisfactory for many years.

Of course, suitable heaters are available commercially, but even a bundle of metal tubes can be used, heated over a gas flame. The heater should be installed in a low cupboard or a screened place so that any bursting of a tube would be enclosed and harmless. A time switch is very useful for an electric heater; decompositions put on overnight are then complete and the tubes cold by next morning.

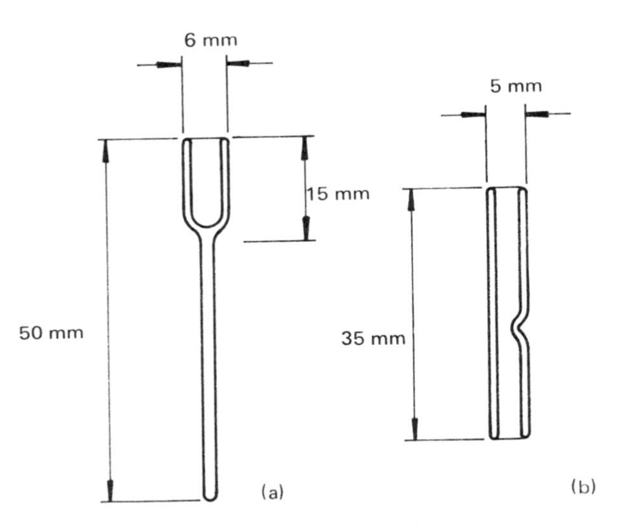

Fig. 3.1 – Weighing tubes (dimensions in mm)

Weighing tubes. Two forms of weighing tubes are recommended.

(a) The form illustrated in Fig. 3.1(a) consists of a glass cup about 15 mm long and of 6 mm outside diameter with a glass-rod handle 35 mm long and about 2 mm diameter. It is best made in soda-glass. It is convenient to provide a small stand and a counterpoise of the same glass, kept always in the balance. Several of the weighing tubes are needed. The long handle serves to keep the sample away from the nitric acid until the tube has been sealed. It also facilitates the removal of silver halide and excess of silver nitrate from the Carius tube in the determination of halogens.

(b) Another form of weighing tube, particularly recommended for phosphorus and sulphur determinations etc., consists of a 35 mm length of Pyrex tubing of about 5 mm outside diameter and 3.3 mm bore. A slight identation is made about 15 mm from one end. In use a wisp of cotton wool (about 1 mg) is packed against the indentation and also in some cases a 50 mg sodium chloride tablet is put in above the cotton wool.

A stand can be made of copper foil to support this tube on the balance pan and a similar tube and stand is used as a counterpoise on the other pan.

This simple tube is particularly easy to rinse out after the Carius destruction.

A safety pipette, 0.5 ml, graduated in 0.01 ml, is used to introduce the nitric acid into the Carius tubes.

A blowpipe for sealing the Carius tubes. An oxy-gas flame of about 10 mm width is used.

Storage block for weighing cups. An aluminium block $2\frac{1}{2}$ in. diameter, $\frac{7}{8}$ in. deep, with 16 holes $\frac{3}{32}$ in. diameter and $\frac{3}{4}$ in. depth, each either numbered or lettered. A recessed flange is machined around the top of the block to take an aluminium cover $2\frac{1}{2}$ in. diameter and $1\frac{5}{8}$ in. high (cut from an aluminium screw-top can.) The clean weighing cups are stored in this covered block until required for use.

Glass stand and counterpoise. A glass tube of about 6 mm outside diameter and about 15 mm deep, having 3 small legs to support a weighing cup on the balance pan. Also a suitable glass counterpoise. Both are kept in the balance used for Carius work.

3.1.2 Reagents

Nitric acid, fuming, s.g. 1.5.
Sodium chloride made into tablets of about 50 mg each (3 mm diameter 3.5 mm long).

3.1.3 Procedure

Keep a number of Carius tubes ready for use. Clean new tubes (to remove any particles adhering to the inside glass surface) with water and a small brush made by winding a pipe cleaner on the end of a long wire. Rinse thoroughly with pure water, leave inverted to drain and dry in an oven at above 100°C. Rinse used tubes thoroughly with pure water, drain and dry in the oven. They can be used repeatedly until they are 150 mm long. (After normal use they can be regarded as exceptionally clean.)

If halogen is to be determined, place about 100 mg of silver nitrate crystals in the bottom of the dry tube, then put 0.3 ml of fuming nitric acid in the tube.

Weigh the sample (which may be from 1 to 40 mg as appropriate) in one of the weighing tubes, containing a sodium chloride pellet when necessary. Place the tube carefully in the Carius tube. It will usually adhere to the inside of the Carius tube because this is moist with nitric acid. With a narrow glass rod push it some of the way down the tube so that the sample is not disturbed and is not heated when the tube is sealed in the blowpipe. Seal the tube in the blowpipe flame. This obviously requires skill, but few find it too difficult after some practice on empty tubes. Aim at a sealed tube of total length no greater than 230 mm.

To seal the tube, first heat it gently in a luminous flame about 200 mm from the closed end. Turning the tube continuously, apply a fairly hot oxy-gas flame to the same position and aim at gathering the glass (rather than pulling it out as soon as it melts) so that the outside diameter remains the same but the glass thickens until the walls are only about 2 mm apart and are about 4 mm thick. Still turning the tube, take it out of the flame and wait a few seconds until the glass begins to solidify. Then pull it out steadily, producing a straight capillary. If this is done well, nowhere is the glass any thinner than the walls of the original tube and a fine capillary is produced which acts as a brake on the escaping gases when the pressure is later released. Using the flame, cut the capillary at a length of about 30 mm, seal its end and anneal all the heated part to some extent in a luminous flame.

When tubes are used again a piece of Carius tubing, open at both ends, is first sealed on and the tube, while still hot, is sealed near the end in a similar way.

When all the tubes of the batch have been sealed and the seal of the last one has cooled, tap the tubes so that the sample tubes reach the bottom, and put them in the heater, in a safe place such as a cupboard, and switch on and arrange the time switch to switch off after about $3\frac{1}{2}$ hr. The temperature will reach about 280°C in about $\frac{1}{2}$ hr and will stay between 280 and 300°C.

If this is done late in the day the tubes are found at room temperature in the morning. Very rarely indeed has an organic substance not been completely decomposed and in all such cases, a smaller sample heated for the same time with the same amount of nitric acid was decomposed successfully.

The process converts all carbon into carbon dioxide and all hydrogen into water. Consequently the sealed tube, even when it has returned to room temperature, is under pressure.

Keeping the tube in the (now cold) heater, expose the tip of the capillary. Release the pressure by applying a flame to the tip of the capillary until the glass melts there and a hole is produced through which the gases escape. Then cut off the end of the tube by applying a very hot glass rod to a file scratch. In some determinations the tube is cut open in a fine blowpipe flame (see under halogen determination, p. 103) [3].

The Carius destruction results in a completely inorganic solution of the element to be determined, in fully oxidized form. This is diluted with water, and contains less than 0.45 g of nitric and nitrous acids. (The latter is easily expelled by warming.) Hence the process is very reliable for the determination of phosphorus, sulphur and various metals.

3.2 OXYGEN-FLASK COMBUSTION

The oxygen-flask technique developed by Schöniger [4] from original work on the macro scale by Hempel in 1892 [5] is a very simple one which requires only inexpensive equipment. It is particularly suitable for samples which are solids or non-volatile liquids, and therefore can be weighed in an open vessel exposed to air. Because of the high temperature reached (~1100°C) and the catalytic effect of platinum, such samples only rarely fail to give complete combustion unless they contain certain metals. The applications of the technique to determination of the halogens and sulphur are described here, and it has been used for several other elements. A number of reviews and a bibliography have appeared in the literature [6–10].

3.2.1 Apparatus

Combustion flasks. Conical flasks with B24/29 ground-glass joints, of 500 ml capacity, are used. Generally, heavy-walled Pyrex iodine flasks with a large collar above the B24 joint are preferred, but ordinary Pyrex flasks are used for the bromine determination, because they have to be heated and cooled rapidly. For the determination of fluorine, the flasks have to be made of quartz instead of Pyrex. For determination of a low level of a component, several portions of sample can be burned successively, with the same absorption solution and flask used.

Platinum sample basket and stopper. This is illustrated in Fig. 3.2. A B24/29 skirted cone (or air-leak) is drawn down beyond the narrower end of the cone to make a tapered part ending about 30 mm from the ground surface. A piece of platinum wire, 1 mm in diameter and 40 mm long, is sealed into this. A platinum basket is formed from a piece of 36-mesh extended platinum mesh, cut to 16 × 24 mm. A silver-steel or stainless-steel rod of diameter 6.4 mm is used to make the basket; this is made easier if a collar, perhaps of brass, is fitted to it 13 mm from one end. The platinum mesh is wrapped round this end, and a seam is permanently welded by hammering the overlapping part (3 mm or so) while it is red hot after heating in a bunsen flame. The 3 mm overlap at the end of the rod

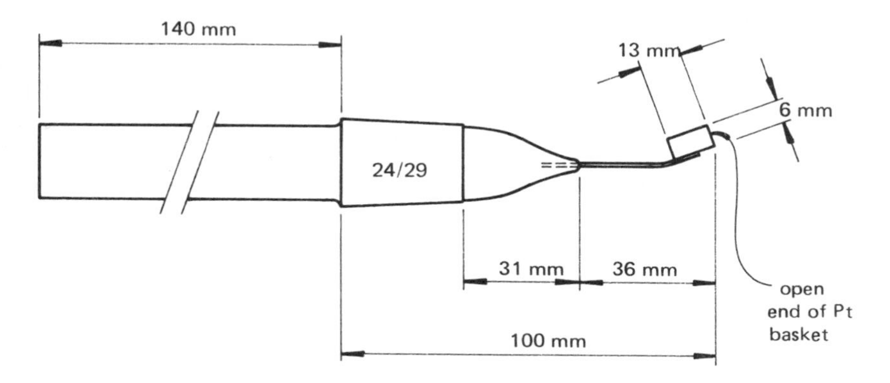

Fig. 3.2 – Platinum sample basket and stopper (dimensions in mm)

is neatly folded to form the base of the basket (three small triangles of mesh are cut out), and the base is completed by hammering while red hot. A 10 mm length of the 1 mm platinum wire fused into the stopper is welded along the side seam of the basket, again by hammering while red hot, so that the open end faces away from the glass stopper. The basket is then about 13 mm long and 6 mm in diameter, and at a distance of about 60 mm from the ground-glass joint.

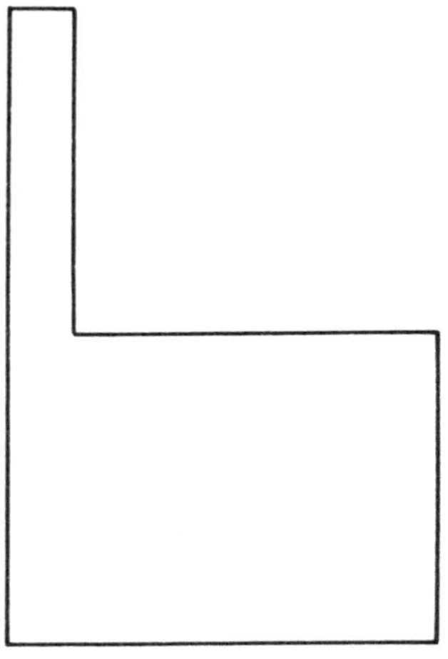

Fig. 3.3

Filter paper 'flags'. These are cut from Whatman No. 1 filter paper to 30 × 40 mm with a tag 30 mm long and 6 mm wide as illustrated in Fig. 3.3. They weigh about 120 mg, including the tag. A metal template, possibly brass, can be used to mark out the filter paper before cutting it.
Oxygen. This can be taken directly from a cylinder of medical-grade oxygen, through a reducing valve, or it may be piped to the whole bench. Near the combustion flask a tap and a glass tube about 20 mm long are needed.
Safety-screen. A very satisfactory screen can be made from a sheet of Makrolon (colourless, transparent plastic) 60 × 40 cm and thickness 3 mm. Two holes are drilled 2 cm from one of the longer edges so that the screen can be suspended on 2 wire hooks and swing freely, yet, with its lower edge about 20 cm from the bench it would screen the operator if an explosion occurred.

3.2.2 Procedure

Weighing of samples. First, rinse the platinum basket and its ground-glass stopper with pure water, and ignite the platinum in a flame to dry it. Take it to the balance.

If the sample is a solid, weigh it in a platinum boat or on a platinum or copper trough (Section 2.3). (If the sample is hygroscopic, this is soon apparent.) During the weighing, fold a filter-paper flag in three, lengthwise, and then unfold it so that it remains creased along the folds. Place it on clean paper (or the notebook) immediately in front of the balance. With forceps, take the boat or trough from the balance pan, and carefully tap the sample, without loss, out onto the central part of the filter-paper flag. Return the boat or trough to the balance pan for reweighing, then immediately fold the paper along the creases to enclose the sample, then roll it and insert it into the platinum basket with the tag outside. Note the weight of the empty boat. (Note that it is *not* satisfactory to weigh the sample with the filter paper on the balance; the weight of the paper may change appreciably.) If the sample is observed to gain in weight while it is being weighed, it may still be worth doing an analysis if the weighing is done as rapidly as possible, but the result will be only an approximate one.

Non-volatile liquids are weighed as follows. Fold a filter-paper flag as before, then cut off the strip representing one third. Fold this strip in the shape of a letter M. Cut a similar strip as counterpoise, then stand both pieces of filter paper in the balance for at least 30 min for their weights to stabilize as far as possible. In all operations, the papers should be handled only with forceps or chamois-gloved fingers. Weigh the M-shaped piece of paper, then place a drop of the liquid sample in the centre part of the M, and reweigh. Lift the paper in forceps, without touching the liquid, enclose the folded M in the remainder of the flag, roll it, and place it in the platinum basket [11].

Alternatively, liquid samples can be weighed by difference from a capillary tube containing a small quantity of the sample. The whole flag is folded into an M-shape, a drop of liquid sample is transferred to the centre of the M by touching the paper with the end of the capillary tube, then the flag is folded and rolled as before and the capillary is reweighed.

Volatile samples are not easy to weigh for oxygen-flask combustion. Combustion of samples sealed into glass capillaries is not usually satisfactory, but it may be possible to weigh a sample into a gelatine capsule, or to seal one into a short length of polythene capillary tubing (the capsule or tubing is wrapped in a flag, and combusted in the usual way). However, it is difficult to obtain capsules and/or tubing with blanks that are reproducible, and low enough. Thus, it is usually better to use an alternative method, if one is available, for such samples.

The combustion. Rinse the combustion flask thoroughly with pure water, and shake out the surplus, then add the appropriate absorption solution (all this may be done before the sample is weighed). Just before doing the combustion, pass a very rapid stream of oxygen (~ 1 litre/min) into the flask for 1 min, then remove the tube.

Light the paper tag, insert the stopper into the flask, and hold the joint tightly closed. At the same time, gently invert the flask so that the absorption solution closes the joint, then hold the flask obliquely so that the hot gases have the maximum distance to travel before they strike the walls of the flask. All this procedure, apart from the lighting of the tag, should take place behind the safety screen. The wearing of strong gloves is recommended by some.

The actual combustion usually takes about 30 sec; no sample or filter paper should remain unburnt. Wait for 1-2 min for the walls of the flask to cool, then shake the still-inverted flask well. Stand it on the bench, add a little water around the outside of the stopper, and leave for at least 5 min for the gases to be absorbed. (In practice, when several determinations are done in series, this time will automatically be longer.) The flask may then be opened, the stopper and platinum rinsed down, and the determination completed in the manner appropriate for the element to be determined.

Safety. The author has never experienced an explosion of an oxygen-flask, although many thousands of combustions have been done in his laboratory. It must be remembered, however, that if the flask were to contain an organic solvent (e.g. acetone or alcohol) an explosion could well occur. For this reason, copious rinsing of the flask with water and leaving it wet with water should be routine, and no solvents should be on the bench where combustions are done. The flask should be filled to overflowing with water during the washing. It has recently been suggested that the wrapping of flasks in Cling Film would hold in glass fragments, and hence reduce the severity of the effects of an explosion, and this very inexpensive precaution is recommended. Electrical ignition, with the flask behind a safety screen, is often recommended [11].

REFERENCES

[1] E. P. Clark, *Semimicro Quantitative Organic Analysis,* Academic Press, New York, 1943.
[2] L. Carius, *Annalen,* 1860, **116**, 1.
[3] J. Unterzaucher, *Mikrochemie,* 1935, **18**, 313.
[4] W. Schöniger, *Mikrochim. Acta,* 1955, 123; 1956, 869.
[5] W. Hempel, *Z. Angew. Chem.,* 1892, **13**, 393.
[6] A. M. G. Macdonald, *Analyst,* 1961, **86**, 3.
[7] W. Schöniger, *Proc. Intern. Symp. Microchem., Birmingham, 1958,* Pergamon, Oxford, 1959, p. 93.
[8] P. Fabre, *Bull. Soc. Chim. France,* 1964, 875.
[9] A. M. G. Macdonald, in *Advances in Analytical Chemistry and Instrumentation,* C. N. Reilley (ed.), Vol. 4., Wiley, New York, 1965, p. 75.
[10] W. Schöniger, *Analysenvorschriften und Literaturnachweise für das Arbeiten mit der Kolben-Verbrennungsapparatur,* Heraeus, Hanau, 1968.
[11] J. P. Dixon, *Modern Methods in Organic Microanalysis,* Van Nostrand, London, 1968, p. 35.

CHAPTER 4

Carbon and hydrogen

4.1 THE BELCHER AND INGRAM EMPTY TUBE METHOD [1]

The method is based on the classical process of complete combustion of the sample in a stream of oxygen, carbon dioxide and water being produced and carried in the oxygen stream into weighed absorption tubes.

Belcher and Ingram [1] in 1950 devised an apparatus which allowed much more rapid combustion than hitherto, and incorporated a specially designed empty combustion tube, containing only silver gauze and a little quartz wool and including a fused-silica baffle-chamber heated at 900°C in which combustion of the sample is completed. About 5 mg of the sample is heated by a movable heater at 800°C in a stream of oxygen flowing at 50 ml/min. Halogens and oxides of sulphur are removed by the packing of silver at 560°C and oxides of nitrogen are removed at room temperature by manganese dioxide in a special tube placed between the water and carbon dioxide absorption tubes.

Flaschenträger absorption tubes [2] (stoppered ends) are used for collecting and weighing the water and carbon dioxide produced by the combustion of the sample. The 'water' tube is filled with granular magnesium perchlorate ('anhydrone') and the 'carbon dioxide' tube with equal lengths of sodium hydroxide on asbestos granules (soda asbestos) and of 'anhydrone'.

Samples containing fluorine, which occur quite frequently in many laboratories today, give no difficulty at all (except possibly perfluoro-compounds, which are difficult to combust), because any silicon tetrafluoride which may be formed can be retained in a section of sodium fluoride granules at 270°C (Belcher and Goulden [3], adapted for this apparatus by Wood [4]).

The author has adopted the practice of using silver granules in the tube instead of silver gauze. This has the advantages that the filling of granules seems to last as long as the tube itself, and the granules can be made from scrap silver.

This method is described here because the author has found it excellent for nearly twenty years. One person can carry out 15 or 16 determinations of carbon and hydrogen in a working day of $7\frac{1}{2}$ hours (two persons with two sets of absorption tubes can do three determinations every hour). It is not difficult to obtain results within 0.1% of the true value for both carbon and hydrogen. Although, as is so often the case with this kind of analysis, results may be better when the apparatus is in continuous use, it is also true that good results can be obtained almost immediately, i.e. after an hour or so, even if the apparatus has been out of use for some time. That is why it is recommended here that it be retained even if automatic C, H & N equipment is also in use, as in the author's laboratory. It is valuable to be able to use an independent method to confirm (or otherwise) results obtained on the automatic apparatus. In addition, residues can be detected and weighed and sometimes metals can be determined, which is difficult with an automatic analyser. Sometimes, also, complete combustion is possible when the automatic analyser has failed to achieve this. If the automatic analyser is giving inconsistent results for a particular sample, incomplete combustion is possibly the reason and good results should be obtained by this older method.

A sample which is hydrated will sometimes, unless special precautions are taken, lose water in the preliminary sweep period in the automatic analysers, giving low hydrogen results. Such samples give no difficulty by the Belcher and Ingram method.

For these reasons, a laboratory which does not have a sufficient work load to justify the purchase of an automatic C, H & N analyser is strongly recommended to use the Belcher and Ingram (1950) apparatus.

It must be pointed out here that the method described later by Ingram [5] in 1961 is probably even better, giving results just as easily and not requiring a short movable burner (resulting in a simpler apparatus). It involves a more rapid ignition of the sample, which is pushed right inside the main heater when the correct moment for combustion is reached. Much of the rest of the apparatus (e.g. absorption tubes) is the same.

With the earlier (1950) apparatus, the author has developed a standardized procedure which is described here. For instance, the absorption tubes are weighed in a strict schedule. In fact, when the hydrogen tube is weighed at the 6th minute it is usually found that its apparent weight is increasing even though it is weighed against a counterpoise of the same dimensions and material, which has been treated in the same way. This may result from this particular tube being nearer to the heaters during the combustion, so that it is still cooling a little at the 6th minute. For very precise results, therefore, the weighing time-table can be extended, say to 10 and 15 minutes for the water and carbon dioxide tubes respectively, but for routine repetitive work this appears to be unnecessary.

If results of the highest precision are required, the blanks should be determined on standard compounds which resemble the sample in question (if sufficient is known about it). For example, a compound with a high nitrogen content may give a high carbon blank, which suggests that a little nitrogen oxide then gets past the manganese dioxide tube.

On one occasion it was found that a highly halogenated compound gave inconsistent results which were later shown to be due to incomplete combustion. Obviously such compounds are often less easy to burn than those containing no halogen. It was found that the temperature of the main heater was lower than it should have been, and results were satisfactory when it was raised again to 900°C. To ensure that the main heater remained at 900°C, a thermocouple was fitted near to the combustion tube and the heating coils were supplied with power through variable transformers. These have been found to be reliable and to require little maintenance. Control by energy regulators ('Simmerstats') was found to be unreliable after long use.

Difficulty has also been experienced with some bromine-containing compounds, as a result of incomplete oxidation and production of some carbon monoxide. The problem can be solved either by using a *slow* stream of oxygen (5 ml/min) in the combustion stage, to increase the residence time of the combustion products in the zone at 900°C, followed by sweep out at 50 ml/min, or by adding an extra pair of tubes to the absorption train, one containing 'Hopcalite' (and anhydrone) to oxidize CO to CO_2, and the other being a second CO_2 absorption tube.

As an example of the determination of a metal, the determination of carbon and hydrogen in a substance containing sodium and gold (the drug, gold sodium thiomalate) can be mentioned. The sample was weighed into a platinum boat and covered with tungstic oxide, and carbon and hydrogen were determined satisfactorily in the usual way. The boat and the various residues were then heated with 30% sodium hydroxide solution, washed with water and dried (or ignited). Then gold alone remained in the boat and could be weighed quantitatively, the tungstic oxide being dissolved away. (The gold alloys to some extent with the platinum boat but this helps to ensure that the irregularly shaped mass of gold remains adhering to the boat during the washing).

When the determination of carbon in some boiler soot was attempted on an automatic C, H & N analyser, low and inconsistent results were found, but satisfactory results (about 90% carbon) were found by use of the Belcher and Ingram apparatus.

Sulphonamides may also cause difficulty, but this can be dealt with by adding vanadium pentoxide as a combustion aid. Compounds which burn explosively in oxygen can sometimes be dealt with by combustion in a stream of air. Various additives have been proposed as combustion aids, for use with compounds difficult to burn completely, and include ignited silver permanganate [6], tricobalt tetroxide [7], vanadium pentoxide [8], tungstic oxide [9], and (for flash combustion) tin(IV) oxide [10]. These additives are also suitable for dealing with compounds containing elements such as B, Si, Se, Ge, Tl, P, metals. Mercury is dealt with by placing gold foil in the beak-unit, after the silver. The foil and the mercury amalgamated with it can be weighed and the mercury determined. Russian workers have developed the use of a combustion tube made in jointed sections, with solid absorbents in various parts, so that several

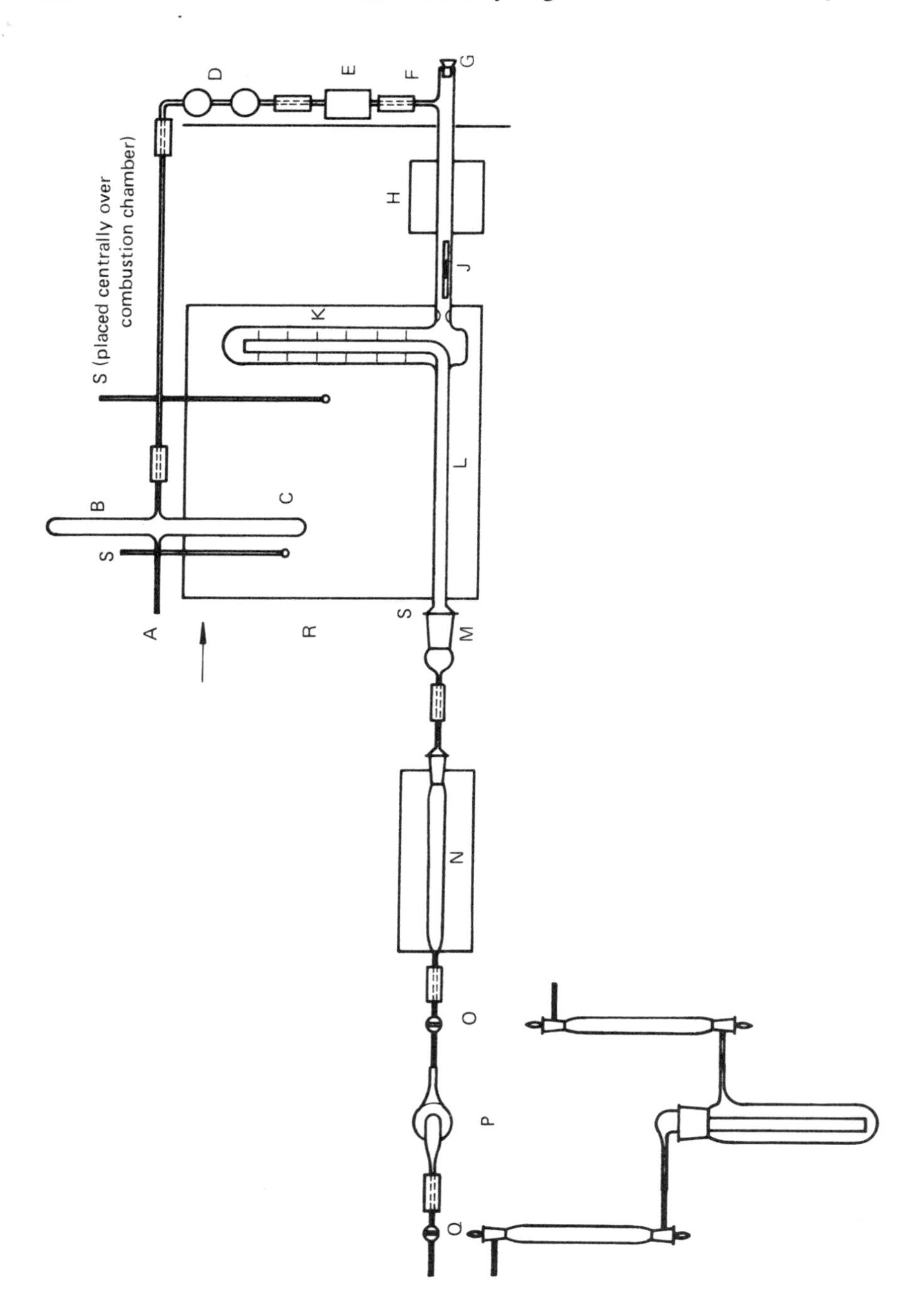

Fig. 4.1 – Diagram of apparatus for the determination of carbon and hydrogen

elements can be determined in a single combustion [10a]. Fluorocarbons are difficult to burn, and combustion aids include silver vanadate [10b] and a sequence of Ag, $3MgO.Al_2O_3$ + Ag, Ag, Pb_2O_3, Ag, PbO_2, Ag in a series of packed tubes [10c].

4.1.1 Apparatus

The apparatus is illustrated in Fig. 4.1. The various parts are as follows.

A. Inlet for oxygen at about 15 psig.
B. Preheater [11] containing platinized asbestos at 600°C for removal of any carbon and hydrogen from the oxygen.
C. Furnace at 600°C for the preheater.
D. U-tube containing soda asbestos and anhydrone.
E. Combined 'Rotameter' flowmeter (with central setting for 50 ml/min) and needle valve for controlling the oxygen flow-rate.
F. Side-arm to combustion tube for entry of oxygen.
G. Well fitting silicone-rubber bung for mouth of combustion tube.
H. Short split-type heater, motor-driven, which can be placed round the combustion tube, giving a temperature of 800°C at the sample boat. A movable Teclu burner is an alternative.
J. Platinum sample boat at centre of cylindrical silica sheath of about 7 mm inside diameter. A platinum wire sealed into a glass rod and terminating in a hook is also required.
K. Carbon and hydrogen combustion tube, baffle chamber heated to 900°C.
L. Granular silver filling heated to 560°C.
M. Pyrex socket fitted to quartz cone of combustion tube and carrying a narrow beak.
N. Sodium fluoride granules in a Pyrex tube heated to 270°C.
O. Absorption tube for collecting water.
P. Manganese dioxide tube.
Q. Absorption tube for collecting carbon dioxide.
R. Casing for heaters.
S. Pyrometers (the thermocouples are placed centrally above and as close as possible to the combustion tube).

Ingram later made the long axis of the heaters horizontal instead of vertical as originally described by Belcher and Ingram in 1950 [1]. This was to prevent the red hot wires from sagging and causing uneven heating and failure.

Belcher and Ingram made their heaters by laying the wire coils in tubular recesses in fire brick (which may easily be drilled). They improved this arrangement still further later by laying the coils in heat-resistant aluminous porcelain tubes (Morgan Crucible Co. Triangle impervious aluminous porcelain, outside diameter 12 mm, and bore 8.5 mm).

The combustion tubes should comply with BS1428: Rapid Method Combustion Tubes (Belcher and Ingram Type) [12].

The Flaschenträger absorption tubes should comply with the standard [12] except that the outside diameter of the main part should be 9 ± 0.5 mm instead of the 8 ± 0.5 mm which is specified. This increases the tube capacity and therefore allows more determinations for every filling. (Incidentally, Flaschenträger himself recommended an outside diameter of 9 mm [2]).

If the capacity (between the ground-glass surfaces) is still found to be less than 5.0 ml it can be increased by heating in the blow pipe and making a slight bulge and slight increase in length. The capacity of the tube is always measured before putting it into use for absorption of carbon dioxide. When not attached to the combustion tube the absorption tubes and counterpoise are carried in a wooden tray with notched sheet-metal supports which give a minimum amount of contact with the tubes.

The Flaschenträger absorption tube recommended for use with the rapid combustion tube has certain advantages over the Pregl-type absorption tube which is described in BS1428 [11].

The tube should satisfy the following requirements:

1. It should be made of soda-lime glass to minimize the risk of electrostatic charge: articles made from borosilicate glass are more likely to develop such a charge.

2. It should be as light as is practicable. The maximum weight when empty is specified as 9 g.

3. It is essential that it contains a sufficient amount and length of granular reagent to retain completely the carbon dioxide or water present in the oxygen stream which emerges from the combustion tube, at a rate which may be as high as 100 ml/minute. For this reason the bore (8.5 mm minimum) and the length between stoppers (106 ± 2 mm) are made as large as is practicable, while still allowing the weight of a filled tube to be low enough to be well within the capacity of a microchemical balance (the filled tube weighs less than 12 g). The tube should also have sufficient capacity to be used many times without the need to renew the reagent.

4. The tube should have as regular a surface as possible. The purpose of the grooves in the stoppers is to confine the grease used for lubrication so that none can be wiped off. The stopper wall, however, should not be too thick (maximum 1 mm) or the total weight will be too great.

5. The stoppers are closed each time the tubes are detached from the combustion tube for weighing. Therefore they should not only be well ground to give a gas-tight fit (when greased) but should not bind or become difficult to turn during use. For the same reason the handle should be large enough to be easily gripped with finger and thumb. It is essential that the hole in the stopper is aligned sufficiently well with the side-arm to prevent grease blocking the latter completely and stopping the gas flow. Well-ground stoppers help to prevent this difficulty by requiring very little grease. The coloured spots are useful for indicating readily the inlet and outlet positions of the tube and showing the side

of the stopper carrying the hole, in addition to showing to which end of the tube each stopper belongs. The coloured spots can be replaced by etched numerals, if that is more convenient. The stoppers are individually ground, so they should not be put in the wrong sockets.

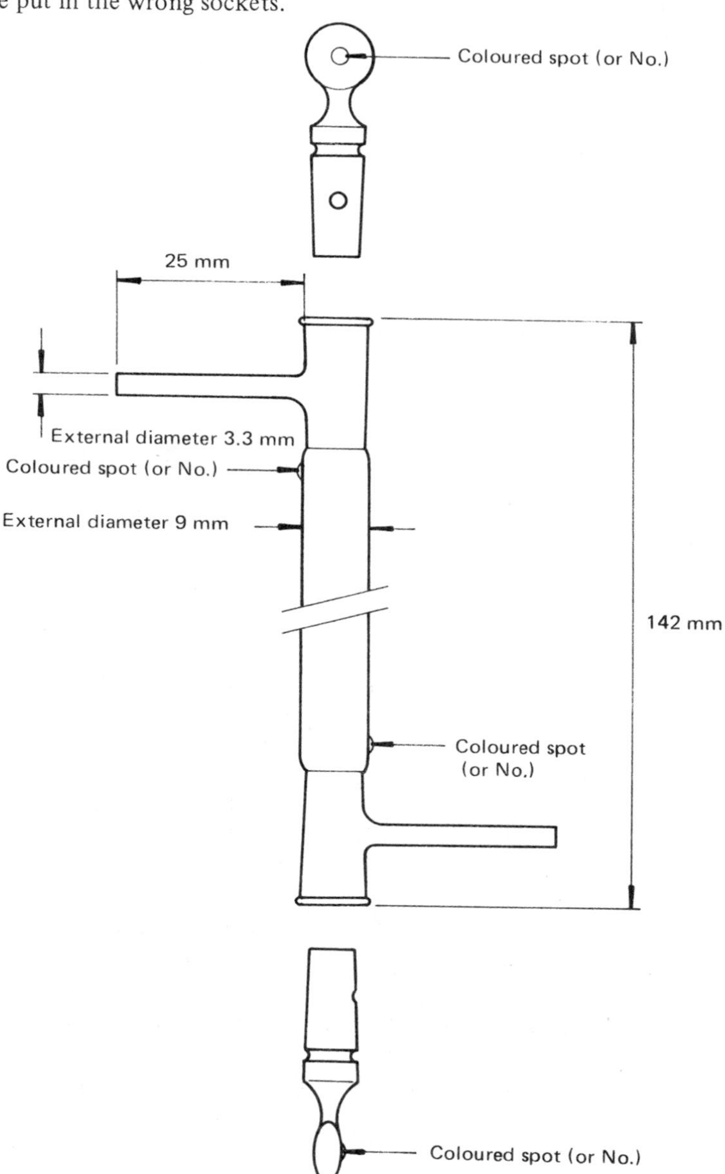

Fig. 4.2 – Flaschenträger absorption tube (dimensions in mm)

6. The side-arms have the same external diameter as the beak (Fig. 4.1) of the combustion tube, to which they are connected by means of flexible tubing. The side-arms have a bore sufficiently large to facilitate cleaning without making them too fragile.

Flexible tubing for connections. All connections are made with a suitable silicone rubber tubing of 3.0 mm bore and 2.0 mm wall thickness.

The tubing is used for the connections between the absorption tubes, but here care must be taken to ensure gas-tight joints. (The lowest permitted diameter of the side-arms of the absorption tube is 3.1 mm and if the diameter is as small as this a good joint with tubing of 3.0 mm bore is not likely.)

The black neoprene tubing specified in BS1428 [11] for connecting absorption tubes, which has a bore of 2 mm and a wall of 3.5 mm has been found to be unsatisfactory because it adheres to the glass surfaces and causes frequent breakages because it is difficult to remove. However, it does have its good points, and it is possible to use it. If the bore is first moistened with glycerine, and care is taken that the glycerine does not affect the weight of the absorption tube, it becomes easier to remove after a few days of use and will then last for many months.

Silica sheaths. Pieces of silica tubing about 50 mm long. They must slide easily into the combustion tube, so the outside diameter should be about 8 mm. The platinum boat must slide easily into them, so the bore should be about 7 mm. Silica tubing of this diameter is ordered in metre lengths and cut as required.

Heater for sodium fluoride tube. This is similar to that described by Wood [4] and is made from a cylinder of aluminium, $2\frac{1}{2}$ in. in diameter and $5\frac{1}{2}$ in. (140 mm) long. A flat surface is made along its length, about $1\frac{3}{4}$ in. (45 mm) in width. An aluminium plate, $\frac{1}{4}$ in. thick, is screwed on to this, holding against the main part a heating element made from Brightray C tape (gauge 0.6×0.1 mm) insulated with mica sheet. Current is supplied by a 0–260 V variable transformer. A setting of about 130 V gives the requisite temperature (270°C). The temperature is shown by a thermocouple placed near the surface of the sodium fluoride tube at about its centre. The heater is drilled axially with a bore of $\frac{1}{2}$ in. (12.7 mm) which easily accommodates the tube of 11 mm outside diameter, containing granular sodium fluoride.

Sodium fluoride tube (N in Fig. 4.1). This tube, (as described by Wood [4]) is made from Pyrex. The main part has an outside diameter of 11 mm, and has a B7 socket at one end, a 30 mm length of beak (3.5 mm diameter) at the other end and an overall length of 190 mm. The stopper consists of a B7 cone and 30 mm tube of 3.5 mm outside diameter.

The movable electric heater (H in Fig. 4.1). The element is a coil of 19-gauge Brightway C wire wound on a $\frac{1}{4}$ in. mandrel, the coil being about $1\frac{1}{2}$ in. long when tightly wound. (Make 20 turns on the mandrel and leave about 3 in. of straight wire at each end.)

In use, it is supported by silica rod or tube inside the coil. It has been found that one heater is sufficient, placed underneath the combustion tube (instead of two, one above and one below as previously described).

Because of the heavy current, the connections to this heating coil should be of heavy gauge. The heaters are supplied with polished aluminium shells which minimize heat losses by reflection, but the shells are better packed with fire-brick (Gibbons HT-1) so that re-polishing the inside of the shells is not necessary.

Current is supplied through a variable transformer used full on, although a variable control is fitted (about 10 V at 20 A). These heavy gauge coils run at such a high temperature that they fail from time to time, but they are easily replaced.

Platinum boats. Although platinum combustion boats complying with BS1428 [13] may be used, provided they will go into the silica sheath, it is more economical to make them as follows:

(a) cut a 25 × 10 mm piece of platinum foil (thickness 0.025 mm);

(b) fold up the walls about 3 mm high all round, doubling at the corners to make a tray about 19 × 4 mm with 3 mm walls. Use the doubled triangular corners as handles at one end, and fold flat at the other.

Counterpoise for absorption tube. The weight of an absorption tube, including stoppers, is adjusted with glass beads and 2 silica wool plugs to be just less than the lighter of a pair of newly filled absorption tubes (usually the water tube).

The stoppers are then fixed in place with Krönig cement, with the glass beads held centrally inside the absorption tube by means of the two silica wool plugs. As the carbon dioxide tube is usually the heavier, an additional counter-poise made of non-magnetic nichrome wire is used when this tube is weighed.

Silicone rubber bungs. The diameter at the narrow end should be $\frac{3}{8}$ or $\frac{13}{32}$ in.

Silica wool plug. The silica wool plug described by Belcher and Ingram [1] is often driven into the combustion chamber if the sample explodes. To prevent this, combustion tubes are now made with a constriction [12] but this is often insufficient. A very small irregular Y-piece made from narrow silica tubing put against the constriction will prevent the silica wool plug from moving into the combustion chamber.

Purification U-tube. Instead of the arrangement described by Belcher and Ingram [1] for purification of the oxygen a specially large U-tube is advisable. The two limbs are filled with soda asbestos and anhydrone respectively and are about 100 mm long and 25 mm in diameter, so that one filling lasts for a long time.

Oxygen regulators. The combustion apparatus often takes its oxygen supply from a copper tube which supplies the whole bench with oxygen, and is connected to the oxygen regulator by a metal union. The regulator has two gauges, the first showing the total pressure (2000 psig for a full cylinder), and the other showing the outlet pressure. The reducing valve is usually adjusted to produce about 15 psig in the bench supply line, or directly to the combustion-apparatus needle valve and Rotameter flowmeter.

Following the purification tubes (D in Fig. 4.1) is fitted a combined Rotameter and needle valve (E in Fig. 4.1). The Rotameter is a simple type, about 70 mm high with a central graduation at 50 ml/min.

Manganese dioxide tube (B.S.1428 [12]).

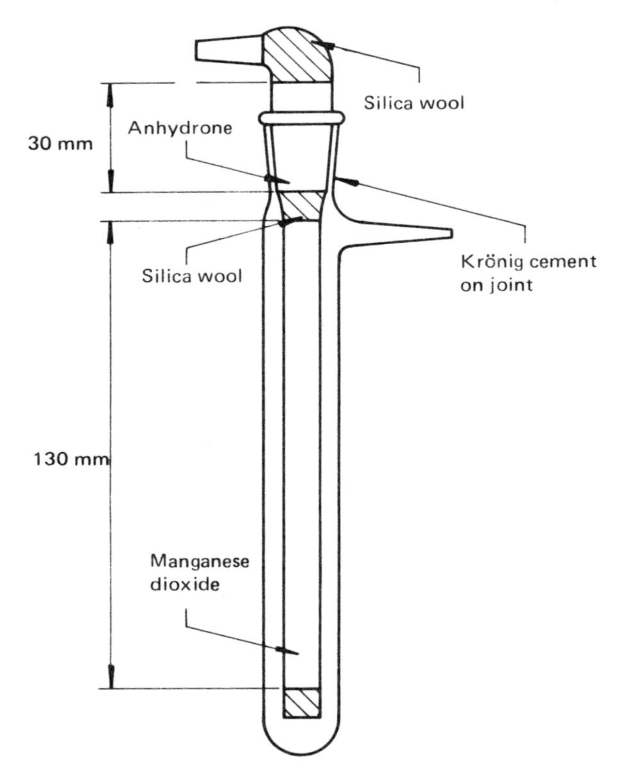

Fig. 4.3 – Manganese dioxide tube

The manganese dioxide tube is illustrated in Fig. 4.3. The side-arms of the tube are stoppered when not in use, except between those determinations which follow each other immediately (i.e. leave open during a normal working day, but stopper overnight and over lunchtime). A Flaschenträger tube may be used instead.

4.1.2 Reagents

Silver wool. M.A.R. Portions are cut off with a knife or razor blade and rolled gently into a plug which can be gently pressed into place in the combustion tube.
Silver granules [14]. Dissolve 200 g of silver nitrate in 2 litres of water. Dissolve 100 g of sodium hydroxide pellets in 400 ml of water and cool. Add this alkali to the silver nitrate solution with stirring until no more silver hydroxide is

precipitated. Heat for 1 hr at 90–95°C (on a steam-bath). Wash the precipitate with water by decantation until the washings are neutral to red litmus paper. Collect the precipitate on filter paper on a Buchner funnel and dry at about 100°C to form a cake. Chop up the dried cake with a razor blade and separate granules between 10 and 22 mesh. Heat portions in a Pyrex boiling-tube (6 × 1 in.) in the muffle furnace at 500–550°C until no black oxide remains (a few min). Wash the granules repeatedly with 1*M* nitric acid, and then by decantation with water till neutral and dry at 100°C for a short time. Store in a bottle kept in a desiccator.

Anhydrone. (anhydrous magnesium perchlorate), M.A.R. 14–22 mesh.
Soda asbestos. M.A.R. 14–22 mesh.
Manganese dioxide, M.A.R. 14–22 mesh, is re-sieved (10–20 mesh) and stored over magnesium perchlorate in a desiccator before use.
Oxygen, medical grade.
Silica Wool. Obtained from Thermal Syndicate Ltd. The grade that is finest in fibre diameter (Grade A) is ordered.
Tungstic oxide. A portion is heated in a muffle furnace (500–900°C) or over a gas flame in a porcelain crucible. It is stored in the muffle or over phosphorus pentoxide in a desiccator.
Sodium fluoride granules [4]. A stiff paste made from sodium fluoride and pure water is dried in the oven at 110°C so that a cake is formed. The cake is cut up carefully and granules of 10–14 mesh are collected on a sieve. These granules, dried again at 100°C, are used for filling the tube (which is heated to 270°C in use).
Apiezon grease L.
Krönig cement.

4.1.3 Filling of Combustion and Absorption Tubes

Combustion tube. The filling of the combustion tube is illustrated in Fig. 4.4.

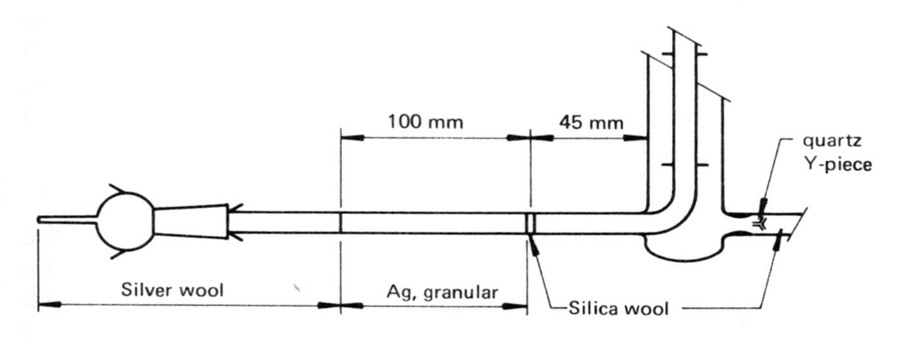

Fig. 4.4

The end and the beak are packed with silver wool, and about a dozen strands of the silver wool are drawn through the beak and cut off flush with its end (to conduct heat and prevent condensation of water).
Absorption tubes. (The CO_2 absorption tube is completely refilled daily when it is in continuous use.) The water absorption tube is filled with magnesium perchlorate and each end is plugged with a plug of silica wool. The carbon dioxide absorption tube is filled half with soda asbestos and half with magnesium perchlorate. Silica-wool plugs are used. Apiezon L grease is used for all the stopcocks. Two pieces of soft chamois leather, about 12 × 20 cm, are needed for cleaning and handling these tubes. Chamois gloves may be worn instead, but must be kept clean.

4.1.4 Procedure

Check that the heaters are at the temperatures specified and switch on the current to the movable electric heater. Refill the carbon dioxide absorption tube (if it has been used for a full day's work). Put an unweighed sample of about 7 mg in the platinum boat. Check on the Rotameter that the oxygen is flowing at a very low rate (the normal stand-by speed). Handling the absorption tubes only with the chamois leathers, attach the carbon dioxide tube to the exit side-arm of the manganese dioxide tube with the magnesium perchlorate section at the exit (upper) end. Then attach the exit end of the water absorption tube to the inlet side-arm of the manganese dioxide tube and finally attach the inlet side-arm of the water absorption tube to the beak of the combustion tube.

Remove the bung from the other end of the combustion tube, with the platinum wire draw the silica sheath containing the used platinum boat to the mouth of the combustion tube, remove the used platinum boat by holding the silica sheath with tweezers and tipping the boat on to the top of the combustion-tube heater, replace the sheath in the mouth of the tube and place the new boat and unweighed sample into the centre of the sheath and then push the sheath with the platinum wire so that the platinum boat will be at the centre of the short heater in its final position (boat about 20 mm from the main heater).

Replace the bung in the mouth of the combustion tube, ensuring a gas-tight fit, open the taps of the absorption tubes in sequence, starting with that nearest the combustion tube and raise the oxygen flow-rate to 50 ml/min by means of the needle valve. Put the movable heater, now at full heat, around the combustion tube and at a distance of 20 mm from the sample boat, switch on the motor and set an alarm clock for 10 min. The movable burner will pass over the sample and reach the main furnace in 4 min and the combustion will be completed and all carbon dioxide and water conveyed to the absorption tubes in the total of 10 min.

During this 10 min period first put the previously used boat on the metal block for a few min and then weigh it. Compare the weight with the previous weight so that it is known if the previous sample left a residue. Weigh out the next sample (5–8 mg) into the boat.

After the 10 min combustion, remove the movable burner from around the combustion tube and boat and turn the oxygen flow-rate down to nearly zero.

Close the absorption tube taps in the reverse order to that in which they were opened. Detach them, touching them with chamois leathers only, and lay them beside the balance on their special tray, with the leathers. At this moment start a timing clock. Weigh the absorption tubes against their counterpoise according to the following routine. Check and if necessary reset the balance zero. While doing this, wipe the side-arms of the absorption tubes. Open momentarily the stopper of the water tube, at the exit end, and with leather-faced tongs place it on the balance hooks. Similarly place the counterpoise on the hooks of the right-hand balance pan. Weigh this tube at the end of 6 min. Similarly weigh the carbon dioxide tube at the end of 9 min, after having momentarily opened the stopper. Then start the combustion of the first sample, already weighed by the procedure described above.

As soon as this combustion is started, the experienced operator, when working in series, should have time to calculate the previous result and to report it and also to weigh the next sample.

Volatile liquids are weighed in glass capillaries as described on p. 24. The glass capillary is held in a roll of platinum foil, which is pinched in tweezers to hold the capillary so that if the liquid explodes during its combustion the capillary is not thrown out of the silica sheath. When an explosion occurs, provided that the capillary stays in place, the result is still satisfactory because of the rapid flow of oxygen.

Samples containing metals, especially sodium or potassium, or containing phosphorus are covered with tungstic oxide after being weighed into the platinum boat. It is an advantage to keep old boats especially for this purpose in order to keep the normal clean boats for other samples so that unexpected residues can be easily detected. If a phosphorus-containing sample is also a volatile liquid which has to be weighed in a sealed capillary, a boat containing tungstic oxide is placed between it and the main heater.

The carbon and hydrogen content of the sample is calculated from the gain in weight of each absorption tube. Carbon dioxide contains 27.29% of carbon and water contains 11.19% of hydrogen.

It is usually necessary to subtract a blank of about 0.1 mg from the weight of water obtained. This amount is found by carrying out a determination several times on standard samples, and finding what blank value needs to be applied to give theoretical hydrogen values, and taking the average. Sometimes a small blank needs to be applied to the carbon dioxide weight, typically about 0.05 mg of carbon dioxide, but this is not so serious. It can be determined at the same time as the water blank.

4.2 OTHER METHODS

The method just described is a development of the original Pregl method of combustion in oxygen and determination of the carbon dioxide and water produced, which was a descendent of the Liebig method. Several other modifications of the method have been made, both to the combustion technique and to the means of measuring the combustion products. The Ingram flash combustion method has already been mentioned. Russian workers [15] also use the empty tube method, but favour combining it with Friedrich's pre-pyrolysis technique [16] of heating first in a limited supply of oxygen. Merz [17] uses the flash combustion technique, in a system automated so that the sample (in a tin boat) is dropped from a magazine into a vertical combustion tube.

Various methods other than weighing have been proposed for completing the determination. Titrimetric methods are based on a chemical reaction between the combustion products and a suitable reagent. Thus the water can be made to react with an acid chloride, and the hydrochloric acid produced can be collected and titrated [18–20]. It can also be reacted with calcium carbide and the acetylene formed can be measured [21,22]. In another method the water is converted into hydrogen chloride by reaction with heated anhydrous magnesium chloride, the HCl is converted into iodine by reaction with silver oxyiodide and the iodine is collected on metallic silver and weighed [23]. The water can also be reacted with 1,1-carbonyldi-imidazole to produce carbon dioxide, which is then absorbed and titrated [24]. The carbon dioxide can be determined by absorption in alkali (usually barium hydroxide) and titration of the excess of alkali, or by conductivity measurement [25,26], or coulometry [27,28]. Another titrimetric method involves absorption in a dimethylformamide/monoethanolamine mixture and titration with tetrabutylammonium hydroxide in a benzene–toluene–methanol mixture [29]; there are several modifications of this method [30]. Unterzaucher [31] has proposed use of the water-gas reaction, first collecting the water on anhydrous barium chloride, then reducing the carbon dioxide to monoxide by passage over heated carbon, then releasing the water from the absorber by heat and passing the water over the heated carbon. The carbon monoxide is then measured in the same way as in the determination of oxygen.

Various manometric methods have been proposed [30], and gas chromatography has also been used [30], in precursors of the automatic analyser methods (Chapter 6). The manometric method is also used in combination with wet oxidation of the sample with various oxidizing mixtures [32,33], or a titrimetric finish can be used. Photometric methods have also been developed [34,35].

Mass spectrometry has been employed for determining carbon, hydrogen and nitrogen in the combustion products [30] and also for investigating the nature of the combustion processes in empty-tube methods [36–38].

A useful review of methods of final determination of carbon dioxide is given by Kozłowski and Namieśnik [39].

Römer and co-workers [40–43] have developed various methods for rapid automated determination of carbon and hydrogen, mainly for very small amounts. The flash combustion method has been automated by Houde *et al.* [44]. The oxygen-flask combustion has been adapted, and several finishes have been described by Sakla and Shalaby [45].

4.3 ENVIRONMENTAL APPLICATIONS

Present day concern with environmental pollution has led to many studies of determination of total organic carbon in the atmosphere and in water, and several modern analysers for carbon and hydrogen have been adapted for the purpose, e.g., by Merz [46] and by Kozłowski and Namieśnik [47].

REFERENCES

[1] R. Belcher and G. Ingram *Anal. Chim. Acta,* 1950, **4**, 118; G. Ingram, *Mikrochim. Acta,* 1956, 847.
[2] B. Flaschenträger, *Z. Angew. Chem.,* 1926, **39**, 717.
[3] R. Belcher and R. Goulden, *Mikrochemie,* 1951, **36/37**, 679.
[4] P. R. Wood, *Analyst,* 1960, **85**, 764.
[5] G. Ingram, *Analyst,* 1961, **86**, 411.
[6] J. Körbl, *Chem. Ind. London,* 1958, 101.
[7] M. Večeřa, D. Šnobl and L. Synek, *Mikrochim. Acta,* 1958, 9.
[8] H. Roth, *Z. Angew. Chem.,* 1937, **50**, 594.
[9] E. Kissa, *Microchem. J.,* 1957, **1**, 203.
[10] V. Rezl and J. Uhdevá, *Mikrochim. Acta* 1979, II, 349.
[10a] J. A. Kuck (ed.), *Simultaneous Rapid Combustion,* Gordon and Breach, New York, 1964.
[10b] P. B. Olson and R. E. Kolb, *Microchem. J.* 1967, **12**, 117.
[10c] C. A. Rush, S. S. Cruikshank and E. J. H. Rhodes, *Mikrochim. Acta,* 1956, 858.
[11] BS1428: Part A1: 1958.
[12] BS1428: Part A5: 1965.
[13] BS1428: Part I7: 1963.
[14] H. Kozumi, in *Microchemical Techniques,* N. D. Cheronis (ed.), Wiley, New York, 1962, p. 438 (cf. *Anal. Abstr.,* 1963, **10**, 421).
[15] M. O. Korshun and V. A. Klimova, *Zh. Analit. Khim.,* 1947, **2**, 274.
[16] A. Friedrich, *Mikrochemie,* 1931, **9**, 27; 1932, **10**, 329.
[17] W. Merz, *Anal. Chim. Acta,* 1969, **48**, 381.
[18] J. Lindner, *Z. Anal. Chem.,* 1925, **66**, 205.
[19] C. J. van Nieuwenburg, *Mikrochim. Acta,* 1937, **1**, 37.
[20] R. Belcher, J. J. Thompson and T. S. West, *Anal. Chim. Acta,* 1958, **19**, 148.
[21] A. A. Duswalt and W. W. Brandt, *Anal. Chem.,* 1960, **32**, 272.
[22] O. E. Sundberg and O. C. Maresh, *Anal. Chem.,* 1960, **32**, 274.
[23] E. Kozłowski and M. Biziuk, *Mikrochim. Acta,* 1979II, 19.

[24] H. A. Staab, *Angew. Chem.*, 1962, **74**, 407.
[25] H. Malissa, *Mikrochim. Acta,* 1957, 553.
[26] S. Greenfield, *Analyst,* 1960, **85**, 486.
[27] D. C. White, *Talanta,* 1966, **13**, 1303.
[28] B. Metters, B. G. Cooksey and J. M. Ottaway, *Talanta,* 1972, **19**, 1605.
[29] R. J. Jones, P. Gale, P. Hopkins and L. N. Powell, *Analyst,* 1966, **91**, 399.
[30] D. H. Davies, *Talanta,* 1969, **16**, 1055.
[31] J. Unterzaucher, *Mikrochemie,* 1951, **36/37**, 712.
[32] D. D. van Slyke and J. Folch, *J. Biol. Chem.*, 1929, **82**, 45.
[33] A. A. Houghton, *Analyst,* 1945, **70**, 118.
[34] N. Oda, S. Ono and H. Matsumori, *Bunseki Kagaku,* 1969, **18**, 854.
[35] N. Gel'man, *Talanta,* 1969, **16**, 464.
[36] W. Walisch and O. Jaenicke, *Talanta,* 1971, **18**, 165, 175; 1975, **22**, 167.
[37] O. Jaenicke and W. Walisch, *Talanta,* 1975, **22**, 345.
[38] W. Walisch and A. Siewert, *Talanta,* 1975, **22**, 355.
[39] E. Kozłowski and J. Namieśnik, *Mikrochim. Acta,* 1979 **I**, 317.
[40] F. G. Römer, G. W. S. van Osch and B. Griepink, *Mikrochim. Acta,* 1971, 792.
[41] F. G. Römer, G. W. S. van Osch, W. J. Buis and B. Griepink, *Mikrochim. Acta,* 1972, 674.
[42] F. G. Römer, C. J. W. van Ginkel and B. Griepink, *Mikrochim. Acta,* 1973, 957; 1974, 1.
[43] F. G. Römer, P. H. van Rossum and B. Griepink, *Mikrochim. Acta,* 1975 **I**, 337, 345.
[44] M. Houde, J. Champy and R. Furminieux, *Microchem. J.,* 1979, **24**, 300.
[45] A. B. Sakla and A. M. Shalaby, *Microchem. J.,* 1979, **24**, 168.
[46] W. Merz, *GIT Fachz. Lab.,* 1975, **19**, 293.
[47] E. Kozłowski and J. Namieśnik, *Mikrochim. Acta,* 1979 **I**, 345.

CHAPTER 5

Nitrogen

5.1 THE DUMAS METHOD

The principle of this old and well known method is to raise to about red heat a weighed amount of an organic substance, intimately mixed with fine copper oxide, in a tube from which the air has been displaced by pure carbon dioxide. When the conditions are suitable the copper oxide acts as an oxidizing agent and the sample is completely decomposed, carbon appears as carbon dioxide, hydrogen as water vapour and nitrogen as nitrogen gas or gaseous oxides of nitrogen; halogens and oxides of sulphur are retained on the copper oxide or absorbed in the nitrometer. The gases are carried in the stream of carbon dioxide through a tube full of heated copper (usually as gauze or granules) where any oxides of nitrogen are reduced to nitrogen gas. Any carbon monoxide produced is reoxidized to carbon dioxide by passage of the gases through more heated copper oxide.

The gases are bubbled slowly into potassium hydroxide solution which absorbs the carbon dioxide but not the nitrogen gas, which is collected and measured in a graduated tube called a nitrometer or azotometer. The weight of this volume of nitrogen, at the measured temperature and pressure, is calculated, and hence the percentage of nitrogen in the sample is obtained.

It is now recognized that unless special precautions are taken some resistant samples may form resins or polymers, and that the carbon which is not completely oxidized may retain some nitrogen.

The essentials for the process are a source of pure carbon dioxide, a heat-resistant tube, heaters to provide the necessary temperatures and a nitrometer.

This chapter describes one commercial apparatus with which the author has considerable experience, but mention is made of several alternatives.

Normally when a C, H & N analyser is used (see Chapter 6) the Dumas method can be dispensed with.

5.1.1 Discussion

The Dumas method, which yields nitrogen quantitatively from nearly all organic substances, has been of the greatest possible importance in organic chemistry. By comparison, the Kjeldhal method has only a limited scope.

Since Pregl demonstrated that satisfactory precision could be obtained with 3-5 mg samples, developments have been aimed at making the process, which perforce must be performed many times, as rapid as possible. Notable was the method of Zimmermann [1] developed during the second world war and also that of Gustin [2] which resulted in the Coleman Nitrogen Analyser. In 1953 Ingram [3] adapted the rapid combustion apparatus of Belcher and Ingram for nitrogen determination. The ultimate in speed is probably the apparatus of Merz [4] (one determination every 4 min, apart from weighing the samples). At about the same time, of course, many laboratories were abandoning the method altogether since the automatic C, H & N analysers made the Dumas method unnecessary and, because in these analysers the sample was combusted in oxygen, errors due to incomplete combustion of the sample did not arise. (The use of C, H & N analysers may not be appropriate, however, for samples with very low and very high nitrogen contents, the first because of the small signal, the second by possible loss on copper oxide.

Many different sources of pure carbon dioxide have been used in the history of the micro-Dumas method. It has been produced by heating sodium bicarbonate and more often by the action of acid on marble chips in a Kipp's or Tucker apparatus. Later, solid carbon dioxide, which became readily available, was found to give a steady supply of carbon dioxide gas of adequate purity if placed in a vacuum flask with an appropriate pressure-release device.

Later it was found that a steel cylinder of carbon dioxide could be suitably purified by sacrificing some of its contents after freezing most of it (p. 54); later still it became possible to purchase cylinder carbon dioxide of adequate purity (better than 99.99% v/v).

Some laboratories or units may still not be able to purchase C, H & N analysers or may not have a sufficient work load to justify them.

For them the following alternatives to the Coleman Nitrogen Analyser are suggested. (1) A relatively simple apparatus based on that described in BS 1428 [5] (or similar simple set-ups that have been described in the literature). (2) Ingram's rapid combustion procedure [3]. (3) The Merz-Dumas nitrogen method. *Nitrogen combustion train* [5]. This provides the basis of a method by which good results can be obtained, though slowly. Although pure carbon dioxide may still be obtained from a Kipp's or Tucker apparatus (acid on marble) or from solid carbon dioxide in a vacuum flask, sufficiently pure carbon dioxide may now be purchased in a steel cylinder. Heaters are now generally electrical and can be purchased or home-made. The combustion tube is made of transparent fused quartz and can be ordered to BS 1428 Part A2. Copper oxide and copper wire-form granules can be purchased. The nitrometer may conform to BS1428 Part D3 or to improved designs such as that of Müller [6] or of Cropper [7].

It might be preferable to follow a procedure such as that of Swift and Morton [8] or of Charlton [9]. Otter [10] described an even simpler arrangement.

Ingram's rapid combustion procedure. This is fully described in Ingram's paper [3] and in his book [11]. He found that the oxygen used had to be generated electrolytically to ensure adequate purity and to avoid large blanks. Nowadays, however, oxygen low in nitrogen can be purchased in steel cylinders for C, H & N analysers. Ingram removed oxygen and reduced oxides of nitrogen in a heated copper-filled tube. This copper is oxidized rather rapidly and must be regenerated, usually with hydrogen, when half used up, perhaps every 2 or 3 days.

The combustion tube, with its baffle chamber, was of the same design as recommended for the carbon and hydrogen determination except that the tube by which the gases leave the baffle chamber was extended in length from 200 to 300 mm to allow a longer copper-filled tube for the removal of excess of oxygen and the conversion of oxides of nitrogen into nitrogen. This is kept at 650°C by means of a split-type electric furnace. The tube must be filled carefully so that channelling does not occur, because of the vital importance of removing all oxygen, which otherwise would be measured as nitrogen.

An ordinary standard glass nitrometer was used by Ingram, but Gustin's design (part of the Coleman Nitrogen Analyser) would probably be an advantage [2].

The Merz–Dumas nitrogen method. This instrument, made by Heraeus (Hanau), is obtainable in England, and is marketed under the name Micro-Rapid-N. It is rather expensive and its advantages, such as extreme rapidity, should be weighed up in comparison with the purchase of a C, H & N analyser. (In his laboratory Merz preferred to determine C & H separately from N because of his very rapid method.) A useful assessment of the method is given by Brewer [12].

A fully automated model is available; this allows 50 weighed samples to be analysed without supervision. The sample, never less than 3 mg (and therefore weighable on an ordinary microchemical balance), is weighed into an aluminium boat and copper oxide is added to it. At the appropriate moment it is dropped into an oxygen-filled vertical quartz tube, at 1000°C or higher. Combustion is nearly always complete. The gases are swept in a carbon dioxide stream (100 ml/min) through copper oxide at 900°C and a large tube of copper at 550°C, then into a nitrometer similar to the Coleman nitrometer, but after the collection of nitrogen in it, its meniscus is automatically returned to a capillary datum line by means of a piston burette motor which is stopped by a photoelectric device. The volume is read out digitally (or printed in the case of the fully automatic apparatus).

5.1.2 The Coleman Nitrogen Analyser

The instrument is supplied complete except for the accessories listed below. It is manufactured for a 115 V electricity supply, so a suitable transformer is required if the mains supply is 240 V. (Coleman Nitrogen Analysers may be bought second-hand because many users of C, H & N analysers no longer need them.)

5.1.3 Apparatus

Carbon dioxide in steel cylinders, analytical grade (better than 99.99% v/v) fitted with a *regulator and gauges.* The regulator must be of the type which will maintain the pressure on the line during the 40 sec purge period when there is a very rapid flow. If the CO_2 supply is not pure enough, purge it as follows. Pack the cylinder to just below the valve stem in solid CO_2 in an insulated container and leave overnight. Take out, open the valve for about 15 sec and close it; repeat three times at 15 min intervals. If the blank is too high, repeat the whole procedure.

The balance should be able to weigh 3-20 mg of sample with an error not greater than 3 μg. Solid samples are weighed in aluminum boats which weigh less than 10 mg, so the Cahn M10 balance was recommended by the manufacturers, but other balances are suitable.

Aluminium boats are supplied by the manufacturer, but satisfactory boats can be made instead from aluminium foil. Pieces of foil are cut to 8 × 20 mm in size. A 10 cm length of brass rod of diameter $\frac{1}{4}$ in. (6 mm) is given a rounded end. The foil is shaped on this end to give a semi-cylindrical boat of size about 20 × 6 mm with one end rounded and one end open. The aluminium should be touched only with clean chamois leather (as gloves or finger stalls, if desired). The sample is placed in the boat towards the rounded end, this end is placed first into the combustion tube, and the boat and sample are allowed to slide down the combustion tube, which is held in a slanting position. If the boats are made from a roll of aluminium cooking foil, and the outer layer is rejected, they can be assumed to be clean. Several blank determinations will confirm this. If desired, however, a number of boats can be washed thoroughly with water and with acetone, dried in an oven at 150°C and cooled in a desiccator. These home-made boats weigh about 7 mg.

Combustion tubes are supplied by the makers. There are 2 types, the combustion tube proper and the reduction tube (the post-heater tube, as it is called in the maker's handbook, meaning the after-heater where oxides of nitrogen are reduced to nitrogen on the hot copper filling). The post-heater tube is not normally heated above 580°C and therefore may be made of heat-resistant glass (the makers use Vycor glass). The combustion tube must be made from fused quartz since it is normally heated to 850°C and occasionally to higher temperatures. Both tubes may be made from good quality quartz tubing of 10.5–11.5 mm outside diameter and wall thickness about 1.2 mm but not less than 1 mm. The combustion tube may be cut on a glass-blower's slitter (diamond wheel) to 350–354 mm in length. The ends should be carefully flame-polished in an oxygen–gas flame so that the tube will be held firmly by the spring to make an absolutely gas-tight joint against the neoprene washers at each end. Similarly the post-heater tube may be made from the same quartz tubing, but its length should be 270 mm.

Neoprene washers, which provide the gas-tight seals at the end of the combustion and post-heater tubes, are supplied by the makers. Because of their vital importance they should be renewed regularly.

A tube rack, obtainable from the makers, is very desirable when the instrument is used near to its full capacity. Tubes can be stored in it before combustion, and it can also hold hot tubes after removal from the instrument. A wooden test-tube rack is an acceptable substitute.

A platinum sleeve, to be used when vanadium pentoxide is added to the sample, can be made from 0.05 mm platinum foil. A 70 × 40 mm rectangle is cut with scissors, then wrapped round a metal rod of diameter about 8 mm, and a seam 2–3 mm wide is made lengthwise (at which there are finally four thicknesses of platinum) and hammered gently on the rod. The diameter of the sleeve is then adjusted so that it will fit the combustion tube as tightly as possible and yet can slide easily in and out of it. A hole pierced through the seam will make it easier to extract the sleeve with a wire hook. In use, the lower end of the tube is crimped so that copper oxide granules cannot drop out of it but gas can flow through it.

Cooling tongs are obtainable from the makers. These can be filled with ice or with solid carbon dioxide. During the purge period they may be held around the combustion tube to cool it to prevent loss of a volatile sample before the combustion stage begins.

A 150 mm quartz dish is suitable for regenerating copper oxide in a muffle furnace.

A combustion tube may be kept ready for use for the regeneration of spent granular copper.

A muffle furnace, a *barometer* and a *thermometer* should be available for use. To calculate the weight of the volume of nitrogen obtained, the temperature and pressure at which it was measured have to be taken into account.

5.1.4 Reagents

Granular copper oxide is supplied by the makers as 'Cuprox'. Wire-form cupric oxide for microanalysis, 40–60 mesh, can be obtained from other sources. To activate it or to regenerate it after use, it should be washed with warm 1% acetic acid, then water, dried at 110°C, and heated in a quartz dish at 850°C in a muffle furnace. Even better, for the most precise work, it may be heated for 30 min in a combustion tube at about 700°C in a stream of oxygen and then allowed to cool in a stream of carbon dioxide. Finally any powder is removed by a 60-mesh sieve. The acetic acid wash is desirable to remove alkalies and thereby to reduce attack on the quartz combustion tube. It also serves to remove copper salts formed in decomposition of halogen or sulphur compounds.

Granular copper is supplied by the makers as 'Cuprin'. In use it becomes oxidized to cuprous and cupric oxides. It may also retain halogens and sulphur oxides. It may be regenerated after removal from the combustion tube, where it tends to cake, by gently breaking up the cake, sieving (40–60 mesh), then heating it gently with a small flame in a combustion tube through which is passed carbon dioxide which has been bubbled through methanol [2]. The formation of pure copper can be seen and controlled. The copper is allowed to cool in a stream of pure carbon dioxide. Hydrogen or carbon monoxide may be used to reduce the copper oxides to copper but methanol is gentler and gives the best product (the aim is to produce granules which are porous and therefore have a high effective surface area).

Pyrex-glass wool is required and is supplied by the makers. It is used in the ends of the combustion and post-heater tubes where high temperatures are not reached. It is preferable to quartz wool, which may form powder and cause leaks, especially at the seatings of the solenoid valves which follow these tubes in the gas flow.

Silver tungstate granules can be obtained from the makers.

Platinized asbestos (30% platinum) may be ordered from suppliers of platinum.

Mercury for the nitrometer.

Vanadium pentoxide should be heated in a muffle furnace at 450°C before use (it melts at 690°C).

Potassium hydroxide solution is supplied by the makers under the name Causticon. It may also be prepared by putting the contents of two 500 g bottles of analytical-grade potassium hydroxide pellets into a 3 litre round-bottomed flask, and adding 1 litre of pure water. The mixture will become very hot; it is because of this that a strong Pyrex round-bottomed flask is recommended. Keep the mixture in gentle motion until all is dissolved, or a small amount at the bottom may never dissolve. When the solution is cold it will probably be quite clear and ready for use. The solution may, however, be filtered by gravity through an ordinary filter paper. Although the paper swells, filtration is quite satisfactory.

5.1.5 Procedure

Combustion tubes are filled as shown in Fig. 5.1. The following is the routine procedure for one operator when one determination follows another at roughly 8 min intervals.

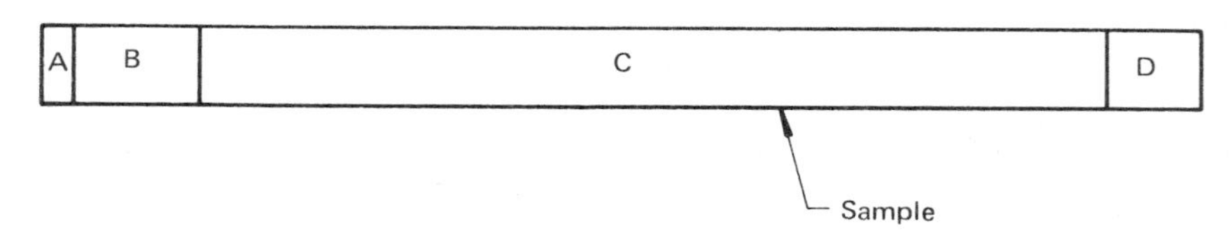

A. Empty 0.25 in
B. Glass wool, 1 in
C. Cupric oxide, wire form
D. Empty, 0.75 in

Fig. 5.1 – Combustion tube

A used combustion tube is allowed to cool, the contents are emptied (as far as the glass-wool plug) into the used cupric oxide bottle, and fresh cupric oxide is added, with tapping, until it is about 13 cm from the top.

An aluminium boat is weighed on the balance, then sufficient of the next sample is weighed into it to produce about 0.6 ml of nitrogen (~ 0.6 mg of N; for example, 10 mg of a 6% N sample). (Volatile liquids are weighed in sealed glass capillaries; these can be weighed on a traditional microchemical balance if the balance used for solids is not suitable.)

The boat containing the sample is allowed to slide gently down the combustion tube until it stops on the granular copper oxide, then the tube is filled with copper oxide to within 20 mm of the end, with gentle tapping. Many substances, when heated in the carbon dioxide atmosphere, will melt and then distil or creep away from the boat, spreading thinly over the copper oxide granules. Complete destruction of the sample with liberation of all its nitrogen can then be expected. (Substances which polymerize or resinify may be expected not to liberate all their nitrogen and the special treatments mentioned below should be tried.)

Fit the combustion tube in place between the spring-loaded clips so that there is a gas-tight joint at each end against the neoprene washers. Set the meniscus of the nitrometer at the starting line, ensuring that the reading on the digital counter is between 200 and 500 μl, and note it.

Turn the combustion cycle control to the start position. The rest of the action is automatic until the volume of nitrogen is read by the operator. First there is a 27 sec pause, then the air in the combustion tube is replaced by carbon dioxide during a 40 sec purge at about 300 ml/min flow-rate. During the last 10 sec of this, the carbon dioxide also passes through the post-heater tube and into the nitrometer. Then the combustion tube is closed at the entry end, and the two movable heaters close around the combustion tube and heat its upper and lower parts to 850°C. Then the two movable heaters move towards each other till they touch, so that the whole tube has been heated and the sample is completely destroyed.

The carbon dioxide flow is automatically started again to sweep the gases through the post-heater tube into the nitrometer where the stirrer ensures that all the carbon dioxide dissolves in the potassium hydroxide solution, and the nitrogen rises to the top, the meniscus of the potash solution dropping accordingly. By turning the knurled ring of the syringe plunger the operator returns the meniscus to the starting point on the nitrometer capillary and the digital counter is read. By subtraction of the original reading on the counter, the volume of nitrogen is found. The temperature and atmospheric pressure should now be noted.

To calculate the percentage of nitrogen in the sample first subtract the blank from the volume found. Then apply the temperature correction (which is needed to allow for expansion or contraction, if the temperatures before and after are different). A table of correction factors appears in the Colemam handbook.

The calculation is based on the fact that one mole of nitrogen (28.016 g) occupies 22.37 litres at STP (760 mmHg and 0°C). Hence 1 ml weighs 1.2505 mg (28.016/22.37) at 760 mm and 0°C. (Tables have been published showing the weights of 1 ml of nitrogen at varying temperatures and pressures [13] and a nomogram relating temperature, pressure and conversion factor from volume to mass is easily constructed.)

Hence the weight of the volume (V ml) of nitrogen found is

$$V \times 1.2505 \times \frac{273}{273+T} \times \frac{P}{760} \text{ mg}$$

(where P is the pressure in mmHg and T is the temperature in °C). The constant terms in this formula, 1.2505 × 273/760 have the value 0.449. This factor is given in the Coleman handbook and may be used for the calculation.

However, the pressure has to be corrected to STP (0°C) and allowance has to be made for the water-vapour pressure above the potassium hydroxide solution in the nitrometer. Tables of correction factors are given in the Coleman handbook. Together, these corrections are almost exactly equivalent to subtracting 11 mmHg from the barometric pressure when the latter is between 740 and 780 mmHg and the temperature is between 25 and 32°C, which in many laboratories is nearly always the case.

Then the percentage of nitrogen in the sample is given by

$$\frac{P_c}{273+T} \times \frac{V_c}{w} \times 0.0449$$

where w is the weight of sample (mg), V_c is the corrected volume of nitrogen (µl), P_c is the corrected pressure (mmHg) and T is the temperature (°C).

Guidance about the various corrections is given in the Coleman handbook; such corrections are an inescapable feature of the Dumas method. After repeated use they cease to become troublesome, but they can be abbreviated by the use of a nomogram or by rough approximations, especially where laboratory temperature varies only slightly. The Pregl-type graduated glass-tube nitrometer also required a correction for incomplete drainage of the alkali. This correction is obviated in the type of nitrometer in which a piston returns the meniscus to the starting point. (The problems caused by the slow attack of the alkali on glass taps are also eliminated.)

To save time and to keep the instrument as ready as possible for use at normal working hours, it is useful to have, in the laboratory, perhaps on a typewritten bench-card, instructions as to what to do at the start and end of the working day and at weekends.

The following would be the text of a typical instruction card (it is assumed that the instrument is in almost continuous use during normal working hours.)

The Coleman Nitrogen Analyser

Regular instructions when the instrument is in use full time from Monday to Friday inclusive.

Monday morning

1. Switch ELECTRICITY ON (WALL SWITCH and LINE SWITCH ON).
2. Turn lower and upper heaters to settings 5 and 6 respectively (or as previously found suitable).
3. Put a FULL combustion tube in place, turn on CO_2 (see below), set the instrument control to SWEEP, and allow the post-heater to warm up by turning its setting to 8 (20–30 min). (The setting may be changed *temporarily to 10* to hasten this process.) Post-heater temperature should be 590°C.
4. Run a complete cycle through PURGE and SWEEP. Then *connect the nitrometer* and measure 2 or 3 complete cycle blanks to ensure that the last one is not more than 20 μl (when corrected for temperature).
5. During any *short* pauses, set the instrument control to SWEEP, with the syringe vent valve open. The next sample may then be combusted WITHOUT further test recycle blanks.
6. Use waiting periods to weigh out samples.
7. After 1 hr break (such as lunch hour) put through 2 complete recycles before combusting samples (need not be measured).

Other mornings

1. During the week arrange for someone to turn heaters up (except post-heater) an hour or more before working time.
2. Deal with post-heater as on Monday morning.

Week nights

1. Turn both lower and upper heaters to setting 2. Leave LINE SWITCH ON. Turn post-heater to OFF. Disconnect the nitrometer and cover both spherical joints.
2. On Friday (at end of week's work) remove combustion tube, cover its lower neoprene washer and switch off electricity at wall switch. Lower and upper heater settings need not be turned down on Fridays.
3. Turn off CO_2 (See below).

Post-heater filling

Renew every 8 weeks (see Fig. 5.2).

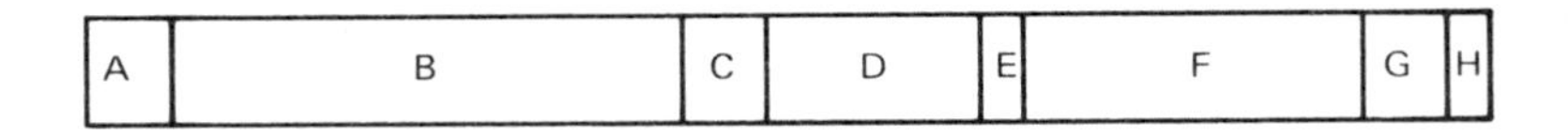

A. Glass wool, 0.5 in
B. Copper granules, 3.1 in
C. Glass wool, 0.5 in
D. Silver vanadate granules, 1.25 in
E. Glass wool.
F. Cupric oxide, 1.9 in
G. Glass wool, 0.5 in
H. Empty, 0.25 in

Fig. 5.2 – Post-heater tube.

Combustion tubes

Dispose of silica combustion tubes after 2 months' use. Retain Cuprox filling for the rewashing process, and the Pyrex-glass wool for re-use. It is useful to number all combustion tubes at their ends (with a diamond) and to note all new combustion tube numbers in a log book.

CO_2 supply

1. Main cylinder gauge should show about 700–900 psig pressure.
2. The gauge on the Coleman Nitrogen analyser should remain at about $4\frac{1}{2}$–5 psig.
3. Each morning the CO_2 is turned on to give this pressure and the stand-by flow of about 5 ml/min is maintained during normal working hours. Each evening at about 5 pm the post-heater is allowed to cool for 15 min with the instrument control set to sweep (i.e. with a flow of CO_2 through the system). The CO_2 is turned off each night. The combustion tube is removed on *Friday evening* only.

Reading the nitrometer

Use the magnet to raise the stirrer bar *3 times* before adjusting the meniscus, and turn on stirrer for a few seconds.

Renew KOH

Tuesday–morning
Thursday–morning
Friday–late afternoon (i.e. refill so as to be ready for Monday).

Post-heater tube cleaning

Empty out 'Cuprox', dissolve the mass of 'Cuprin' in concentrated nitric acid and finally soak the tube in chromic acid cleaning mixture, wash well and dry at 100°C.

Procedure when the result is suspected to be low

If the result for nitrogen is suspected to be low (incomplete liberation of nitrogen) first try mixing a small amount of sample with fine copper oxide and combust with moving heaters at a higher temperature. The blank must be re-determined under the same conditions. If a higher nitrogen content is then found it may be worth repeating the analysis with the samples in the platinum sleeve with vanadium pentoxide added [14]. Weigh the sample in an aluminium boat and put 50 mg of vanadium pentoxide over it. Put 5–7 mm of copper oxide in the platinum sleeve against the crimped end and then insert the aluminum boat containing the sample and vanadium pentoxide. Slide the sleeve gently into the tilted combustion tube and then fill the combustion tube copper oxide to within $\frac{1}{2}$ in. of the top in the usual way.

The addition of the following reagents has also been recommended: nickel sesquioxide, cobalt oxide, potassium dichromate, copper acetate, potassium chlorate. About twice the sample weight should be used in each case. The blank should be known in each case because most of the reagents provide oxygen and its complete removal afterwards must be ensured.

In a busy laboratory, where the automatic nitrogen analyser is fully occupied, too many variations in technique cannot be tolerated so the first procedure given above is recommended. It should be pointed out that suspicion that the apparatus is giving low results usually arises when the other results (C,H,halogen,S) agree with those expected, within the usual experimental error. However, even then it is by no means certain that the nitrogen determination is faulty. A Kjeldahl nitrogen determination would really resolve the problem. It is not unknown for a research worker to be completely mistaken about the reactions he has postulated, and for his product not to have the expected composition. If the product is impure, the number averages for the other constituents may by chance correspond to the expected true value.

5.2 THE KJELDAHL METHOD

Sufficient sample to contain about 1 mg of nitrogen is heated with concentrated sulphuric acid, potassium sulphate, and small amounts of mercuric sulphate, selenium, sucrose and cigarette paper (used to make cups for weighing solids).

The mixture is heated so that it just boils. The purpose of the potassium sulphate is to raise the boiling point of the concentrated sulphuric acid. With suitable samples all the nitrogen is converted into ammonium sulphate and organic compounds are destroyed, the carbon being removed as carbon dioxide.

Heating is continued for a total of about 1 hr to ensure that all the nitrogen is converted into ammonium sulphate. When cool, the mixture is diluted with water, made alkaline with sodium hydroxide solution containing thiosulphate, and the ammonia is steam-distilled, collected and titrated with 0.01*M* hydrochloric acid. The sodium hydroxide solution contains thiosulphate in order to produce insoluble mercuric sulphide during the distillation, to prevent ammonia being held back as mercury ammine complexes.

5.2.1 Discussion

There are numerous variations of the micro-Kjeldahl determination of nitrogen. The method described here has proved to be very rapid and practical. One person can carry out about 18 determinations in a normal working day. The only types of compound with which the method has been known to fail (that is, give low results) are those in which one or more of the nitrogen atoms is attached to an oxygen atom or to another nitrogen atom. Some of these (e.g. nitro compounds) can be converted quantitatively into ammonia if the sample is first reduced with hydriodic acid. Baker [15] found that nitro-groups could be made to yield all their nitrogen as ammonia by treatment at 430°C in a sealed tube with sulphuric acid in the presence of mercuric oxide as catalyst and thiosalicylic acid or glucose to reduce the nitro-groups.

Both the hydriodic-acid treatment and the sealed-tube digestion are extra complications which detract from the simplicity of the ordinary direct Kjeldahl procedure, and if they are necessary it is often preferable to determine the nitrogen by another means such as a modified Dumas method.

Nevertheless, it is often valuable to have in the laboratory a completely independent method for the determination of nitrogen, as a confirmatory method. It has also been found that the Kjeldahl method can give very precise results, even when the nitrogen content is in the lower ranges (5% or lower) provided quite large samples (40–50 mg) are available.

The ammonia distillation apparatus of Schöniger and Haack [16] is the best of the many that the present author has used. In the method described no acid is put in the receiver into which the ammonia is distilled [17] (hydrochloric, sulphuric or boric acids are often recommended). It has been found that in the condenser spiral, which is over a metre in length, the condensed water becomes cold well before it reaches the bottom and therefore it retains all the ammonia. This arrangement is better than a short condenser which dips for only a centimetre or two into acid in a flask. If boric acid were used to absorb the ammonia, the end-point of the titration would be less sharp because of the buffering effect of the boric acid.

An alternative arrangement is to put into the receiver a volume of 0.01*M* hydrochloric acid that is not quite enough to neutralize all the ammonia, and to complete the titration with further 0.01*M* acid. No other standard solution is then needed for the titration.

However, if the nitrogen content is lower than expected, the acid in the receiver is not all neutralized and the determination has to be completed with

0.01*M* sodium hydroxide. This method has been found to give very precise results.

The electric immersion heater with bare metal wire suggested by Clark [18] works very well, giving rapid responses, but a gas flame is also quite satisfactory and nearly as rapid.

If the catalyst mixture can be made into tablets in large quantities, there is a considerable saving of time and blank values are consistent.

5.2.2 Apparatus

Digestion flasks (30 ml nominal volume). These are described in BS1428 [19]. They have an overall length of about 165 mm, a neck of outside diameter 18 mm, and a pear-shaped bulb of 40 mm diameter and a capacity of approximately 30 ml. A labelling badge is placed half-way up the neck so that it faces the viewer when the spout is to the left.

Burette for concentrated sulphuric acid. This can be made in the laboratory from tubing of outside diameter about 15 mm, with graduations about 17 mm apart, i.e. corresponding to 2 ml intervals. It has a glass tap at the bottom with its key arranged vertically so that it cannot drop out, and which is always lubricated with the concentrated sulphuric acid itself. The top has a ground-glass joint carrying a trap containing sulphuric acid to prevent entry of ammonia from the atmosphere.

Rack for digestion flasks. This consists of a wooden bar, notched to take the flask necks, supported by a wooden frame, and with asbestos board on the base to carry six flasks (which are very hot immediately after the digestion).

Draining rack for digestion flasks. A stainless-steel wire-mesh rack which will not let pass the 40 mm bulb of the digestion flask. After rinsing, the flasks are inverted in the rack to drain ready for re-use, and nothing is put inside the flask (such as draining board pegs).

Digestion apparatus. This has places for six flasks, each of which can be heated over an individually controlled gas flame. The exhaust manifold is connected to a water-pump (preferably all-glass) to carry away most of the acid fumes. The digestion apparatus is fully described in BS 1428: Part BI: 1964 (Clause 3 and Fig. 1) [4]. Electric heaters are not recommended because they need to have separate controls for each flask and may be expensive.

Distillation apparatus. This is illustrated in Fig. 5.3. A description of the main part was given by Schöniger and Haack [16].

Cigarette paper cups. 'Blue' cigarette papers are used. Circles are cut out round a sixpence (19.5 mm diam) and are then formed into a cup round the flattened end of a glass rod of 8 mm diam.

Glass cups, 7 mm in diameter, are made by closing the end of suitable soda-glass tubing to make a flat bottom, and cutting the end off to make a cup about 6 mm high.

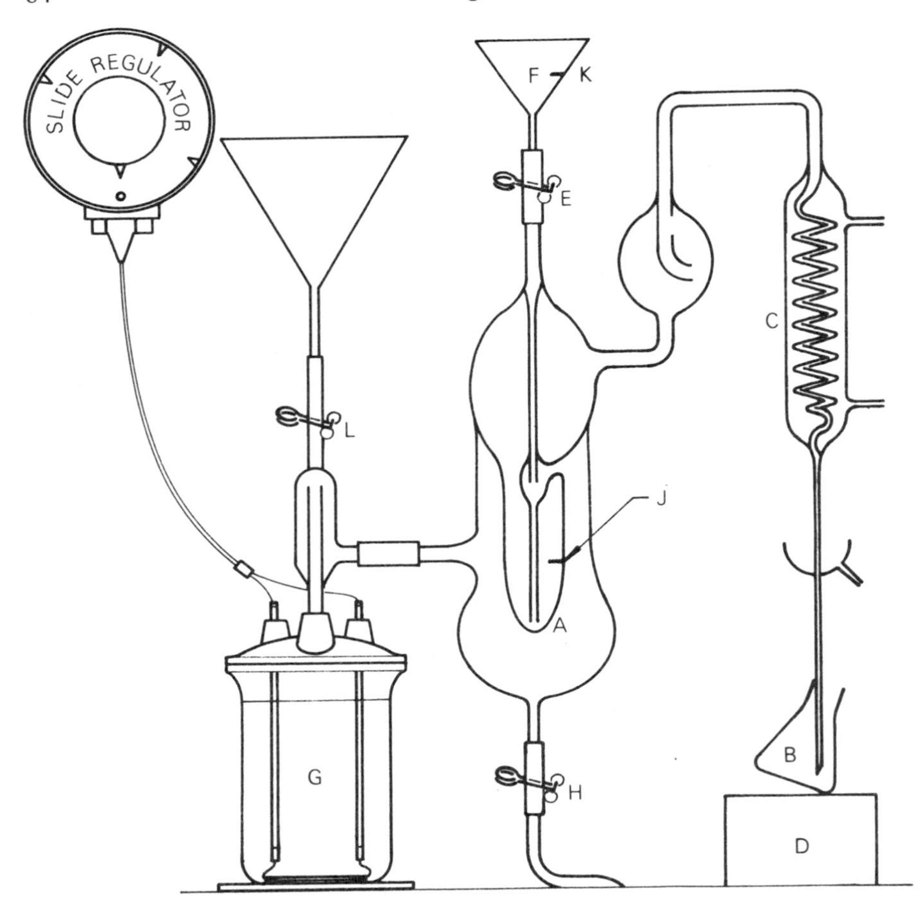

Fig. 5.3 – Distillation unit

Sealed glass bulbs. Bulbs about 4 mm in diameter are blown in the end of standard melting point tubes. Then a capillary end of about 20 mm is drawn out.
Receiver flasks. These are 50 ml Pyrex conical flasks.
Steam generator (Fig. 5.4). Steam for the ammonia distillation is generated in a 2 litre Witt pot containing a coil of wire (resistance 62 Ω) immersed in distilled water. The heating rate is continuously variable as the current is supplied to the wire from a 5 A variable transformer. Three marks are made on the dial of the transformer to correspond to the following settings.
(a) 'Simmer' to keep the water just boiling when not in use – 90 V (130 W).
(b) 'Rapid Heat' to bring to the boil as quickly as possible – 210 V (706 W).
(c) 'Standard Boiling Rate' (for the distillation proper, distilling about 20 ml of water in 6 min – 155 V (384 W).

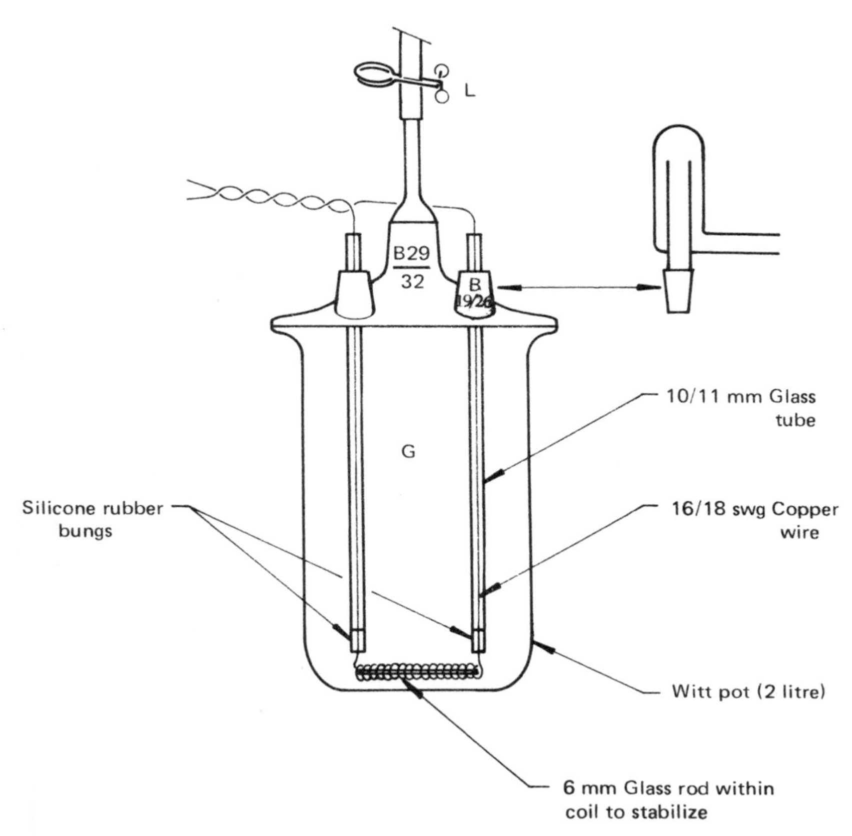

Fig. 5.4 – Steam generator

(With 250 V applied, the heating power is 1 kW, and will bring the full boiler to boiling point in about 12 min).

The copper leads (16 s.w.g.) are passed through glass tubing of 10–11 mm outside diameter sealed with a silicone-rubber bung at both ends. The upper ends of the tube enter the Witt pot through one-hole No. 17 rubber bungs.

Heating coil. A single coil of resistance 62.5 Ω is formed by tightly winding 23.7 ft of 26 s.w.g. Kanthal wire onto a $\frac{1}{4}$ in. steel rod. The coil so formed is pulled out to give about 30% extension. Each end of the coil is tightly wound three times round the clean end of a 12 in. length of 16 s.w.g. copper wire and silver-soldered, with Easiflow flux. Excess of flux is removed by leaving the joints in boiling water or a steam-bath for 5–10 min. After the copper leads have been fitted through the silicone-rubber bungs, leaving about 1 in. of copper wire projecting, the exposed copper wire and silver-soldered joint may be covered with 'Araldite' or other epoxy cement to prevent galvanic corrosion between the Kanthal wire and the copper wire.

Burette, 10 ml, graduated at 0.02 ml intervals, with automatic filling device, for 0.01*M* hydrochloric acid.

5.2.3 Reagents

Potassium sulphate. Use a grade which is already powdered: do not use M.A.R. grade because this is in large crystals and needs grinding. The small nitrogen content (if any) is determined and allowed for in the overall blank.

Mercuric sulphate

Selenium powder

Sucrose

Boric acid

Catalyst tablets. Make up a mixture of 4 kg of potassium sulphate, 624 g of mercuric sulphate and 250 g of sucrose. Turn in a clean ball-mill for 5 hr. Remove the balls, then add 124 g of selenium and mix by turning for 20 min (this is to avoid contaminating the stone balls and jar with selenium). Give the mixture to the tablet maker with instructions to granulate by adding 150 ml of 60% sucrose syrup (i.e. a further 90 g of sucrose in 150 ml of water).

After granulation, add 150 g of boric acid powder (about 3% to lubricate the tablet die; keep away from ammonia fumes or the blank may be unduly high). Make tablets weighing ~1 g each and of diameter ½ in. (13 mm). Over 5000 will be obtained.

Sulphuric acid. Use a low in nitrogen grade. The nitrogen content will be included in the blank, but if 2.5 litre bottles are used, it is not necessary to determine the blank very frequently.

Distilled and demineralized water. All water used (pure water) is distilled and demineralized in the permanent laboratory apparatus, except that, for the steam boiler, distilled water is always used (not demineralized water).

Solution of sodium hydroxide (40% w/w) and sodium thiosulphate (5% w/w). Empty two 500 g bottles of sodium hydroxide pellets into a 2 litre round-bottomed flask and add 1 litre of pure water. Rotate gently until all the sodium hydroxide has dissolved (the solution becomes hot, but if it is left to stand too soon it will cool and complete dissolution will be impossible). Dissolve 125 g of sodium thiosulphate pentahydrate in 375 ml of pure water in a 2 litre conical flask with gentle warming if necessary. Allow both solutions to cool overnight, then mix them and store in a Winchester. There is slight warming on mixing. The thiosulphate is added solely to precipitate mercuric sulphide on boiling and to liberate ammonia from any mercury–ammonia complexes that may be formed from the mercury used as a catalyst in the digestion.

Methyl Red solution. Shake 1 g of Methyl Red with 200 ml of ethanol (industrial spirit). Leave to stand until as much as possible has dissolved. Decant a portion into a dropping bottle.

Methylene Blue solution. Dissolve about 0.1 g of Methylene Blue in 100 ml of pure water.

Hydrochloric acid, 0.01M. Take 500 ml of standard 0.1000*M* hydrochloric acid (a commercial standard solution is convenient) in a dry standard flask and empty it and rinse it several times with pure water into a 5 litre container and make up to the mark with pure water. This solution will be sufficiently close to exactly 0.01*M* for practical purposes, but in any case divergences will be corrected for by use of an empirical factor in the calculation (note that it is often assumed that 1 ml = 0.142 mg of N instead of the theoretical 0.1401 mg of N).
Hydriodic acid, S.G. 1.7, M.A.R. This is supplied in 5-ml sealed glass ampoules.
Hypophosphorous acid, 30%.
Red phosphorus, amorphous.

5.2.4 Procedure

Digestion. Into a digestion flask which has been rinsed and allowed to drain in the rack, place a catalyst tablet and roughly 2 ml of conc. sulphuric acid from the special burette, then add the weighed sample.

Weigh solids and viscous non-volatile liquids (syrups, oil and gums) in cigarette-paper cups. Counterpoise the cup on the left-hand pan of the balance by a similar one on the right-hand pan, weigh, add the sample and weigh again. A liquid sample can be dropped into the cup from a spatula. As much as 50 mg of sample may be taken when the nitrogen content is low. Otherwise the amount taken should give a titration of 8–10 ml. For nitrogen contents over 25% take about 5 mg, although the titration will be greater than 10 ml, and the burette will have to be refilled during the titration. Drop the paper cup and sample carefully into the digestion flask (already containing the tablet and acid). If a liquid sample soaks through the paper, stand one cup on a second slightly flattened one (cup and saucer), and use two cups as counterpoise.

If a liquid is found to be slightly volatile, weigh it in a glass cup. Drop the sample and cup into the digestion flask. The glass cup tends to cause bumping during the digestion and its use should therefore be avoided if possible. Slightly volatile liquids give good results by this method because the liquid contained in the glass cup has a relatively small surface area if a sufficiently large sample is taken (so loss by evaporation is negligible, especially if it is dropped into the acid immediately after weighing).

Weigh very volatile liquids in sealed glass bulbs which are fragile enough to be crushed with a glass rod under the surface of the acid in the digestion flask. Weigh the bulb empty, fill it by applying a vacuum when the open end of the capillary is under the surface of the liquid, release the vacuum, centrifuge, seal and weigh again. Drop it into the digestion flask, crush with a clean glass rod and rinse the rod into the flask, before removal, with a little pure water.

With the water-pump on, to remove fumes, heat the flask on the digestion apparatus in the fume cupboard over a small flame for 3–4 min, then with a nearly full flame so that the mixture finally boils and a ring of sulphuric acid can be seen as the reflux boundary in the neck of the flask. Before this stage is reached the mixture darkens and fumes are given off. Some of the fumes are carried off by the water-pump. The dark mixture turns pale after a few minutes. Heat strongly for a further 40 min from this stage, then turn off the flame and allow the mixture to cool to room temperature. Solid potassium bisulphate will separate.

During the digestion, add distilled water to the boiler of the distillation apparatus G, (Fig. 5.3) and turn the heater on so that the water is boiling by the time the digestion is completed.

Pretreatment with hydriodic acid (the so-called HI-Kjeldahl). Nitro compounds may sometimes be converted into amino compounds and thus rendered suitable for the Kjeldahl method by treatment with hydriodic acid. This is performed as follows. Weigh out the sample in the normal way but drop it into an empty digestion flask (which need not be quite dry). Add 15 drops of hydriodic acid and warm gently over a steam-bath. Add 2 drops of hypophosphorous acid until all the sample is in almost colourless solution. If need be, add more hypophosphorous acid and hydriodic acid, dropwise with heating, until this is achieved but without exceeding 30 drops of hydriodic acid. Add about 10 mg of red phosphorus to minimize bumping, then reflux gently over a small flame on the digestion stand for 15 min. Cool and add about 6 ml of pure water (bulb half-full). Cool again and add 2 ml of conc. sulphuric acid from the special burette. Boil gently until nearly all the iodine is removed, then cool and add about 2 ml of water. Boil gently until this water is removed and white fumes appear, then add a catalyst tablet and boil gently, increasing the flame gradually to avoid frothing. Continue as for an ordinary determination, but boil for only 30 min after clearing (instead of the usual 40 min).

Distillation and titration of ammonia. Turn on the condenser water of the distillation apparatus, C in Fig. 5.3. Add about 20 ml of pure water to the distillation vessel (A) and boil out the apparatus, as for a normal distillation (see below), for 6 min, rinsing as described below. On switching off, the apparatus empties itself. Rinse with four funnelfuls of water to cool the apparatus and use immediately, in the same way that one distillation follows another in routine analysis.

Meanwhile, dilute the contents of each flask with pure water so that each bulb is about half full, and allow to cool slowly before proceeding.

Add to a receiver flask (B), 10 ml of pure water, 3 drops of Methyl Red solution and 1 or 2 drops of Methylene Blue solution. As soon as the distillation apparatus is ready, position the receiver, by means of a wooden block (D), tightly up under the condenser in an oblique position. Pour the contents of the digestion flask through the funnel (F) with the pinch-clip, E, removed, and follow with four thorough rinses with pure water (but do not fill above the 35-ml mark on the distillation vessel at J). Close the funnel inlet with the pinch-clip E and add about 7 ml of 40% sodium hydroxide solution (to the measuring mark on the funnel K). Run in the alkali solution by opening the pinch-clip, E, then close the pinch-clip again, and fill funnel F with water. Switch the boiler to rapid heat and close pinch-clip L on the boiler to direct steam into the distillation vessel A.

Set an alarm clock to ring after 4 min. As soon as the mixture in the distillation vessel A is boiling vigorously, turn the transformer control down to standard boiling rate. This will give about 20 ml of distillate in 6 min.

During the distillation, titrate the previous distillate, if any.

Prepare the receiver for the next charge by rinsing the flask just used for the titration, filling it etc., as above.

When the alarm rings, lower the receiver to the bench and set the alarm clock again for 2 min. The water distilling will then rinse the condenser tip for 2 min. At the end of this 2 min (i.e. a total of 6 min distillation) empty and rinse the distillation vessel as follows. Switch off the current to the boiler (mains switch). As the contents are sucked out of the distillation vessel, rinse it four times with pure water from the funnel, controlling the pinch-clip so that as little suction as possible is lost. (Note that this not only rinses the vessel but also cools it, so that the next charge does not heat up prematurely). Run the effluent to waste by opening the lower pinch-clip (H). Open the pinch-clip, E, to the filling funnel, F. Open the pinch-clip to the boiler, L. (Put the clip on the stem of the filling funnel.) Turn the water-heater control to simmer after switching it on. Set aside the receiver with the distillate and place the next receiver under the condenser and proceed with the next distillation, during which the titration is done.

Calculate the nitrogen content of the sample as follows.

$$\% \mathrm{N} = \frac{(T-b) \times 14.0}{W}$$

where T is the titration volume (ml), b is the blank titration (ml) and W is the weight of sample (mg).

Each ml of 0.01M hydrochloric acid = 0.14 mg of N.

If desired, an empirical conversion factor can be determined by titration of distillates from a series of samples of pure standard substances.

To determine the blank, carry out at least six determinations on one sample, e.g. atropine, using progressively less of the sample (e.g. 20, 15, 10, 7, 5, 3, 2 mg). Plot titration volume against weight of sample. Extrapolate to zero sample weight to obtain the blank, b ml. (This is normally 0.1-0.2 ml of 0.01M acid).

5.3 OTHER METHODS

Neither the Dumas nor the Kjeldahl method is particularly suitable for very low levels of nitrogen. The Kjeldahl digestion and distillation can be used for such samples, however, if a different final determination method is used.

The Nessler reagent [20] or the chlorophenol-indophenol reaction [21] can be used. The distillation can be omitted and the ammonium salt in the acid digest determined with ninhydrin [22]. If no preliminary reduction has been necessary, the acid digest can also be analysed by making it alkaline, oxidizing the ammonia with hypochlorite [23, 24] or hypobromite [25, 26] added in known amount, and titrating the surplus oxidant.

REFERENCES

[1] W. Zimmermann, *Mikrochemie,* 1944, **31**, 42.
[2] G. M. Gustin, *Microchem. J.,* 1960, **4**, 43.
[3] G. Ingram, *Mikrochemie,* 1953, 131.
[4] W. Merz, *Z. Anal. Chem.,* 1968, **237**, 272.
[5] BS1428: A2: 1959.
[6] A. Müller, *Mikrochemie,* 1947, **33**, 192.
[7] F. R. Cropper, *Analyst,* 1954, **79**, 178.
[8] H. Swift and E. S. Morton, *Analyst,* 1952, **77**, 392.
[9] F. E. Charlton, *Analyst,* 1957, **82**, 643.
[10] I. Otter, *Nature,* 1958, **182**, 656.
[11] G. Ingram, *Methods of Organic Elemental Analysis,* Chapman & Hall, London, 1962.
[12] R. Belcher (ed.) *Instrumental Organic Elemental Analysis,* Academic Press, London, 1977.
[13] H. Gysel, *Prozenttabellen organischer Verbindungen,* Burkhauser-Verlag, Basel, 1951.
[14] P. D. Sternglanz and H. Kollig, *Anal. Chem.,* 1962, **34**, 544.
[15] P. R. W. Baker, *Analyst,* 1955, **80**, 481.
[16] W. Schöniger and A. Haack, *Mikrochim. Acta,* 1956, 1369.
[17] J. Vene, *Bull. Soc. Sci. Bretagne,* 1938, **15**, 49.
[18] E. P. Clark, *Semimicro Quantitative Organic Analysis,* Academic Press, New York, 1943.
[19] BS1428: Part BI: 1964.
[20] W. G. Frankenburg, A. M. Gottscho, S. Kissinger, D. Bender and M. Ehrlich, *Anal. Chem.,* 1953, **25**, 1784.
[21] E. D. Noble, *Anal. Chem.,* 1955, **27**, 1413.
[22] S. Jacobs, *Analyst,* 1960, **85**, 257.
[23] R. Belcher and M. K. Bhatty, *Mikrochim. Acta,* 1956, 1183.
[24] M. Ashraf, M. K. Bhatty and R. A. Shah, *Anal. Chim. Acta,* 1961, **25**, 448.
[25] H. H. Willard and W. E. Cake, *J. Am. Chem. Soc.,* 1920, **42**, 2646.
[26] H. W. Harvey, *Analyst,* 1951, **76**, 657.

CHAPTER 6

Automatic C, H & N analysers

When Pregl introduced micro methods for the determination of carbon, hydrogen and nitrogen in organic substances the advantage lay not only in the small amounts of sample required but also in the greater speed of determination. Since then, research in organic chemistry, with the synthesis and extraction of new organic compounds, has increased exponentially and more rapid methods have become increasingly desirable. Increased speed has generally been achieved by making the equipment more complicated (and more expensive), a sad negation of Pregl's simple methods, but it cannot now be denied that the modern automatic analyser has filled a need and has proved its worth. It is interesting that such analysers were developed just after the use of gas-liquid chromatography became widespread, and modern analysers often incorporate separation on a column and detection by thermal conductivity.

The first instrument available in England for the rapid determination of C, H & N was made by Technicon. In the Technicon method about 0.3 mg of sample was combusted, so a special sub-micro balance was needed. The combustion took place in helium containing 3% of oxygen and the products of combustion, including nitrogen, were measured by integrating the areas of peaks detected by katharometers and registered on recorders.

Later the Perkin-Elmer Model 240, the Hewlett-Packard and the Carlo Erba instrument became available. The last of these is discussed thoroughly in Belcher's book [1], which can be recommended as an invaluable help to any user.

There is now no doubt that these instruments have proved their worth, and they can be recommended whenever the considerable expense is justified. The author will confine himself here to describing the principle on which two of them operate and to giving some hints on their use.

Use of automatic analysers is not confined to determination of carbon, hydrogen and nitrogen. Oxygen or sulphur may also be determined along with these three elements, and other combinations are also possible.

6.1 THE PERKIN-ELMER MODEL 240 C, H & N ANALYSER

Once the day's preliminaries have been carried out, including blank determinations, the routine repetitive operation of the instrument is as follows.

About 2 mg of the sample is weighed accurately on a suitable balance mounted near the instrument. The weighed sample and boat are placed under cover on a metal block until the end of the previous determination is indicated by a light showing on the instrument, then the closure to the end of the combustion tube is removed. The ladle containing the previous sample boat is removed with metal tweezers (gloved hands), the boat tipped out and the new boat and sample put in, the ladle replaced in the combustion tube and the combustion tube closed. The start button of the instrument is pressed and the operator immediately begins to weigh the next sample, usually in the boat which has just been removed, after igniting and cooling it. When indicated by a light on the instrument (this has been changed by some users to an audible signal) the sample is pushed into the hot combustion zone by means of a magnet (there is a magnet in the ladle). The operation is then automatic except that when the recorder comes into action the operator must select suitable attenuation factors for C, H & N by turning knobs. Finally, the signals for C, H & N are read, including the zeros for each. The carbon, hydrogen and nitrogen contents found are then calculated and the results reported. A result is obtained every 13 min. Although everything possible has been made automatic there is usually sufficient for the operator to do to make the work quite interesting and satisfying. After practice there is even some waiting time. Operators do not complain if they have to operate it all day every day for a week or so, but it can be appreciated that an almost automatic balance helps to avoid faulty weighing caused by inattention due to monotony.

The functioning of the instrument is, briefly, as follows. The sample is combusted in pure undiluted oxygen, the products are swept on in helium, and carbon dioxide and water are absorbed in absorption tubes in the traditional manner. The flow of gas is controlled by means of solenoid valves, which open when electrically activated. The first stage is the purging of any air which may have entered with the sample and its replacement with pure dry helium. A metered amount of oxygen is then passed in and the signal given to place the sample in the hot zone (about 950°C). The combustion takes place in a closed compartment. Helium then sweeps the contents of this compartment into a reduction tube containing hot copper granules where any excess of oxygen is removed and any oxides of nitrogen are converted into nitrogen, and then into a mixing bulb followed by a sampling coil. This is controlled by a pressure switch which closes the valve when a certain helium pressure is reached. The measurements are made by three pairs of thermal conductivity cells (katharometers). The mixing bulb, sampling coil and katharometers are all contained in a box at a constant temperature near to 75°C. The sample of gases, which is now of uniform composition, passes through the one cell of the first pair, through an absorption tube which removes water vapour and finally through the other cell of the pair. The resistances of the cells are thus rendered out of balance. An increasing voltage is automatically applied until they are balanced, and is registered as a vertical straight line on the recorder chart. This signal represents the hydrogen content of the sample. Similarly the next pair of cells measures the carbon content by comparing the thermal conductivity of the gases before and after passage through a tube which removes carbon dioxide. Finally the gas sample is compared with the original helium, to find the nitrogen content. In all three cases an arbitrary preset zero is subtracted. The recorder (potentiometer) readings are straight lines (because the gas mixture is not changing in composition, unlike the situation when a peak is produced in gas chromatography).

The ratio of the recorded signal to amount of carbon, hydrogen and nitrogen is determined by weighing and combusting pure standard organic substances.

Each day's work is preceded by one or more blank determinations, combustion of unweighed samples to condition the apparatus, and calibrations. The calibration does appear to change slightly from day to day. However the instrument is capable of continuous use, giving more than 20 determinations of C, H & N every day on samples of unknown composition.

Although other methods have been suggested, it has been found that volatile liquids can be weighed in small glass capillaries as soon as the operator has had sufficient experience to gauge about 2 mg of the sample. Liquids that are not volatile can be weighed into platinum boats in the same way as solids.

The original setting up of the instrument, including fitting of helium and oxygen cylinders and their regulators is described in the Perkin-Elmer manual which also describes the signs that warn when maintenance is required (tube fillings exhausted, valves in need of cleaning). In practice, the experience of the operator in recognizing these signs is invaluable.

It has been found useful to prepare and keep in the laboratory cards with concise instructions for periodical maintenance, as follows.

Reduction tube change (time required – 10 min + 2 cycles).

Fill a reduction tube in advance with 60–100 mesh copper (see diagram). Preserve it ready for use in an evacuated tube with a glass tap closure.

Set programme wheel to 9, and remove furnace protective cover.

Undo nut at right-hand end of reduction tube. Loosen screw on retainer bracket.

Undo nuts on connector fitting (CARE – HOT!). Slide off fitting and remove nuts from combustion and reduction tubes. Remove reduction tube (use tongs).

Slide in new reduction tube with the silica wool plug at the right-hand end.

Burn off any dark oily deposit from the end of combustion tube. Replace the connector with a clean spare. (Clean the used one later by washing with water and acetone.)

Refit connector and nuts, using new O-rings. Tighten nuts carefully. Tighten screw on retainer bracket.

Press START button to complete cycle (the programme wheel may be rotated manually if time is short.) Run TWO blank cycles. Leave helium pressure at 14 psig overnight.

Tighten joints the following morning.

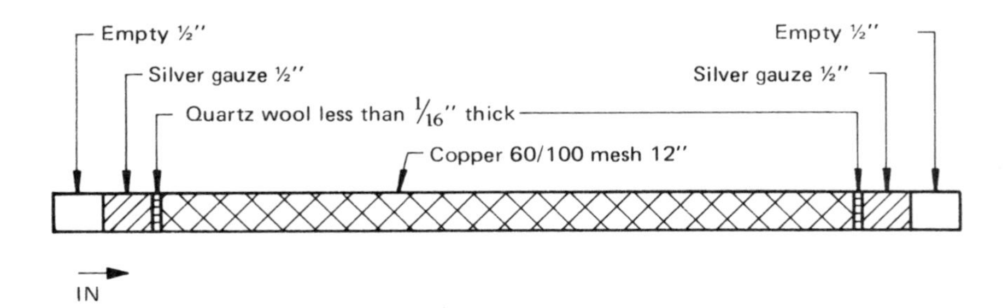

Fig. 6.1 – Reduction tube

Oxygen cylinder change (time required – 80 min + 2 cycles).

Close exit valve on the regulator (clockwise rotation), and close oxygen cylinder valve.

Change oxygen cylinder, then turn on cylinder valve. High-pressure gauge should register 400–420 psig. Open fully the exit valve on the regulator.

Set programme wheel to 2, fit entrance flow plug. Set power switch to DETECT.

BLEED DOWN. Open oxygen cylinder valve for 30 sec. Close firmly and allow pressure on main gauge to fall to zero and to 2 psig on low-pressure gauge. Perform this procedure THREE TIMES (20–25 min when regulator is set for 30–35 psig). Test for leaks during this procedure. Open oxygen cylinder valve after Bleed Down.

Check oxygen flow-rate with soap-bubble flowmeter and adjust to 50 ml/min.

Replace sample entry plug. Turn programme wheel clockwise to 0. Run a minimum of TWO BLANK CYCLES.

N.B. Allow extra time if the helium and oxygen scrubbers are being changed.

Helium cylinder change (time required – 15 min + 2 cycles).

Close the exit valve on the regulator (clockwise rotation) and helium cylinder valve. Set power switch to OFF and programme wheel to 0.

Change helium cylinder. [Carefully note the correct adaptor(s) to use for each type of helium cylinder.] Turn on cylinder valve. High-pressure gauge should register 2000–2500 psig. Open fully the exit valve on the regulator. Test for leaks.

Set power switch to ON and programme wheel to ½. Leave sample entry plug in place.

BLEED DOWN. Open helium cylinder valve for 30 sec, close firmly and allow pressure on both gauges to fall to zero. Perform this procedure FOUR TIMES.

Rotate programme wheel clockwise to 0. Wait 5 min and check that the Red Warning Light is off. Set power switch to DETECT.

Run TWO BLANK CYCLES. If changing oxygen cylinder as well, start doing so before these blank cycles.

Combustion tube change (time required – 15 min + 2 cycles).

Prepare reduction and combustion tubes in advance (see diagram). (A new reduction tube is always fitted at the same time, normally at the end of the day.)

Manually rotate programme wheel to 9. Remove ladle from combustion tube.

Undo nut at each end of reduction and combustion tubes. Loosen screw on retainer bracket. Slide off connector fitting. Inspect the high heat coil and remove tubes (tongs).

Fit new tubes (the silica wool plug on the reduction tube should be on the right. Take care not to damage the high heat coil when fitting the combustion tube. Tube ends should be level at the left-hand ends.

With new O-rings, refit connector and nuts. Tighten screw on retainer bracket. Replace ladle and tighten all nuts.

Press START button to complete cycle. (The programme wheel may be rotated manually if time is short.) Run TWO BLANK CYCLES and leave helium pressure at 14 psig overnight.

Tighten joints the following morning.

N.B. If the high heat coil is very distorted and/or tight-fitting, replace it by removing coil cover, undoing Allen set-screws, fitting replacement, tightening Allen screws and refitting cover. Attend to this during the procedure above.

(Full details on separate card.)

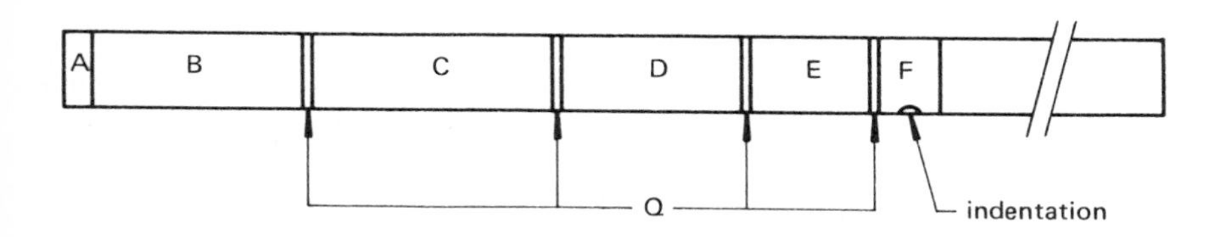

Fig. 6.2 – Combustion tube filling for C, H & N analyser

A. 0.25 in. empty.
B. Silver gauze, 1.75 in. see below.
C. Silver vanadate granules, 2 in.

D. Silver oxide and silver tungstate on 60-80 mesh Chromosorb P, $1\frac{1}{2}$ in.
E. Silver tungstate on magnesium oxide, 1 in.
F. Platinum gauze, 0.5 in. see below.
Q. Quartz wool, less than 2 mm.
Total length 25 in. (635 mm).
B. Silver gauze – cut $6 \times 1\frac{3}{4}$ in. Roll tightly, soak in 'Teepol' for 5 min, and wash well. Heat in flame to dull red heat.
F. Platinum/rhodium gauze – $12 \times \frac{1}{2}$ in. Roll, but not tightly, and insert from the short side of the dimple.*

Water and carbon dioxide traps (time required – 10 min + NO cycles)

Prepare tubes in advance as shown in the diagram (Fig. 6.3).

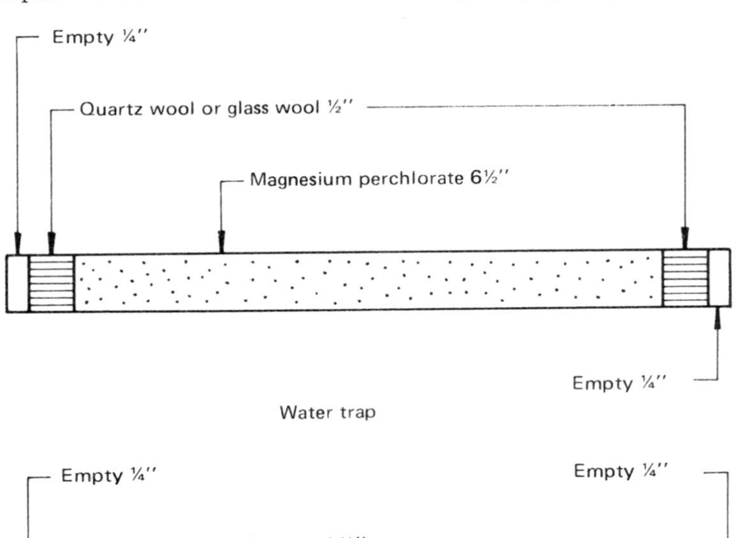

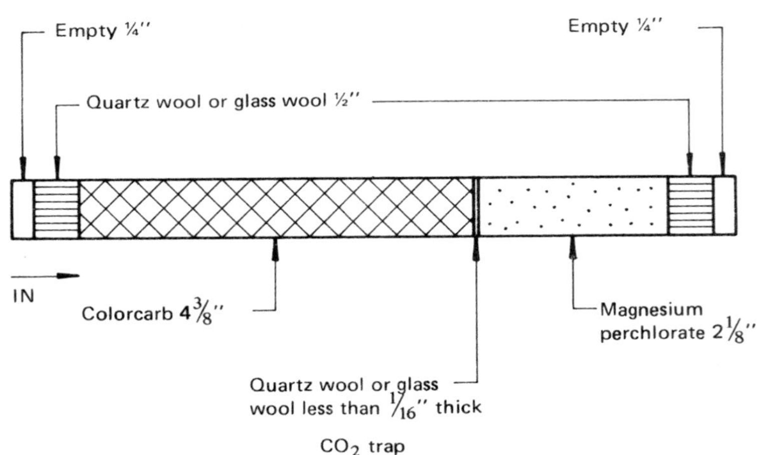

Fig. 6.3

*Grade 4 platinum gauze, 36 mesh, 34 gauge (0.009 in. diameter wire) has been found satisfactory and less likely to crumble after use.

Set programme wheel to 0. Turn power switch from DETECT to ON.

Undo nuts on right-hand side of exhausted tubes, loosen bracket screws. Undo left-hand nuts.

Renew the O-rings only if deformed. Put in fresh traps, tightening left-hand nuts first, move bracket to allow about 1 mm of end-play in the tubes, tighten bracket screws, and finally the right-hand nuts. Ensure that the white magnesium perchlorate filling is on the right-hand side.

Switch out the red light. (Reset button is immediately next to it.)

Turn the power switch from ON to DETECT.

Leave overnight or for at least 3½ hr, then tighten the nuts.

Helium and oxygen scrubbers change (time required – 10 min).

Prepare tubes beforehand as in the diagram (Fig. 6.4).

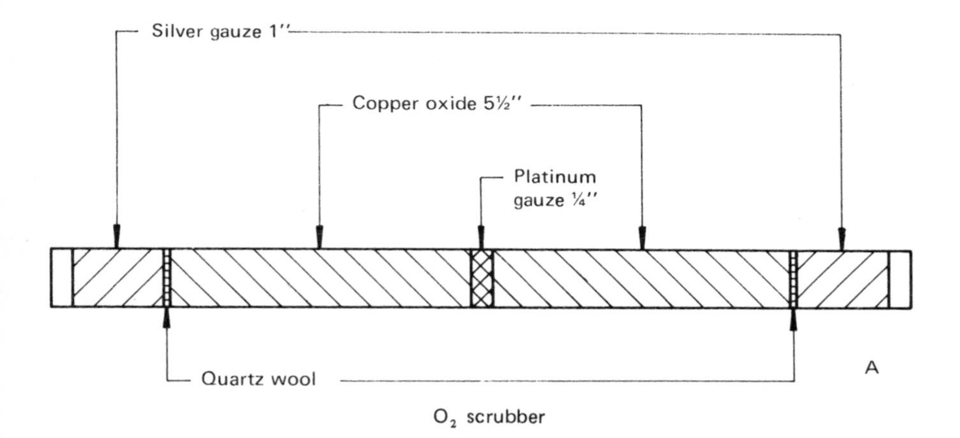

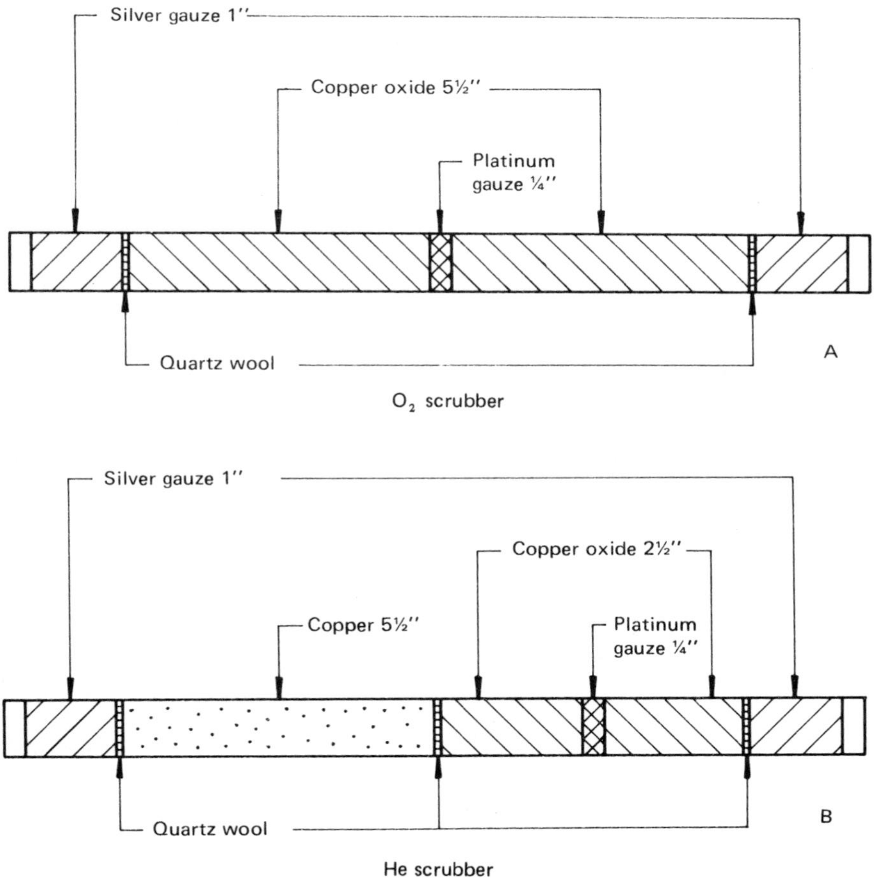

Fig. 6.4

Set programme wheel to 0. Turn power switch from DETECT to ON.

Turn off helium and oxygen cylinder regulator exit valves.

Undo nuts on right-hand side of exhausted scrubbers, loosen bracket screws. Undo left-hand nuts.

Renew the O-rings only if deformed. Put in fresh scrubbers, tightening left-hand nuts first, move bracket to allow about 1 mm of end-play in the tubes, tighten bracket screws and finally right-hand nuts. Ensure the white magnesium perchlorate filling is on the right.

Turn on the helium and oxygen cylinder regulator exit valves.

Turn the power switch from ON to DETECT.

Tighten nuts the following morning.

Solenoid valve – Cleaning and Parts Replacement (time required – 10 min per valve for cleaning, 5 min per valve for parts).

Set programme wheel to 0 and power switch to OFF.

Turn off helium and oxygen cylinder regulator exit valves.

Remove housing nut and washer on sleeve assembly ($\frac{9}{16}$ in. A.F. spanner).

Remove the coil assembly. Place spanner wrench nut over the sleeve assembly and unscrew the sleeve assembly ($\frac{3}{4}$ in. A.F. open-ended spanner).

Clean the sleeve assembly, the plunger spring, the plunger assembly and the O-ring, with methylene chloride. Do NOT immerse the plunger assembly or O-ring in the methylene chloride for longer than 20 sec.

Reassemble in reverse order. Turn on helium and oxygen valves. Turn power switch to ON. Switch out the red light. (The reset button is immediately next to it.)

Turn the power switch from ON to DETECT.

N.B. If it has been off for about 10 min the detector oven will require $\frac{1}{2}$–1 hr to stabilize.

Leaking valves may be detected by removing the upstream side and applying soap solution.

High heat coil change (time required – 15 min + 2 cycles).

Manually rotate the programme wheel to 9.

Slide off the high heat coil shield and carefully remove the combustion tube (*q.v.*).

Loosen the Allen set-screws in the coil supports. Lift out the high heat coil and ceramic bar.

Insert new high heat coil and ceramic bar (renew shield if necessary) and tighten screws.

Carefully replace new/old combustion tube, renew any distorted O-rings and tighten joints.

Press START button to complete cycle (programme wheel may be rotated manually if time is short). Run at least TWO BLANK CYCLES.

6.2 THE CARLO ERBA ANALYSER, MODELS 1102 AND 1104

The Carlo Erba instrument includes a permanent channel for the determination of oxygen in addition to the channel for simultaneous determination of carbon, hydrogen and nitrogen.

In the later model 1106, the oxygen channel can be converted to allow the determination of sulphur.

The routine operation of the instrument for C, H & N determination is as follows. About 0.6-1 mg of sample is weighed into a cylindrical tin cup of diameter 4 mm and height 6 mm, made of tin foil 0.025 mm thick. (Suggested balances include the Sartorius 4125 and the Mettler 22.) The top, empty part of the tin cylinder containing the weighed sample is flattened with tweezers and folded over twice. The capsule is then folded again in the other direction to form a small pellet, care being taken not to split the foil. The sample in the tin foil is then put in a numbered, 5 mm diameter compartment of a circular magazine. Weighing is continued until there is a sample in all 23 compartments. Volatile liquids are weighed into small glass tubes sealed by a special technique (p. 86). Several samples of standard compounds are included in the 23. (One or two may contain no nitrogen, as a check on the nitrogen blank reading.)

After about 30 min of preliminaries, such as purging with helium, the first cycle is started with no sample, the 2nd cycle and particularly the combustion of the sample is observed to ensure it appears normal and then the combustion of all 23 samples is allowed to proceed. This takes 4-5 hr, and does not need the attendance of an operator, so it can go on out of working hours. For each sample, numerical quantities corresponding to the CO_2, H_2O and N_2 produced will be printed by the integrator. The signals are derived from the areas of the peaks produced as the gases are carried by a helium stream into katharometers after chromatographic separation. Factors are calculated from the results for the standard compounds, and used to calculate the carbon, hydrogen and nitrogen contents of the unknowns. An electronic calculator is helpful for these calculations, or a computer may be used.

The design of this intrument is extremely complex, but the analysis proceeds, briefly, as follows. At intervals of about 10 or 12 min a sample is dropped from the magazine into a vertical combustion tube at 1050°C. Just beforehand, the helium in this tube is enriched with oxygen so that a vigorous combustion takes place. Heat is also generated by the oxidation of the tin container. The combustion products are swept by helium through the lower part of the combustion tube, which contains hot granular chromic oxide followed by silvered cobalto-cobaltic oxide. Thus carbon dioxide and water are produced from the sample. The gases pass into a second tube containing silvered copper granules at 600-650°C, where excess of oxygen is removed and oxides of nitrogen are converted into nitrogen gas. Halogens and oxides of sulphur are retained on hot silver.

The carbon dioxide, water, and nitrogen are completely separated on a chromatographic column (Porapak QS,) at 120°C, and are detected by thermal conductivity (i.e. in a katharometer). The peaks produced are integrated automatically and the corresponding figure is printed out. These figures, as mentioned above, are related to amounts of carbon, hydrogen and nitrogen by calibration with pure standard compounds.

Recently it has been found that the hydrogen (i.e. water) blank is more consistent if the helium is passed through a tube of copper sulphate pentahydrate to humidify it before it enters the apparatus. This eliminates the problems caused by adsorption of water on very dry parts of the gas train when changing from samples of very low (say 1%) to very high (11%) hydrogen.

The instrument works best in a laboratory which has temperature control to ± 1°C. Greater variations (e.g. 5°C in an hour) can cause the factors to drift, and recalibration is necessary.

The accumulation of debris from the tin capsules must be removed after 200–300 cycles (perhaps weekly).

Clearly the method does not permit determination of any ash due to the presence of metal or silica (perhaps from chromatographic purification).

6.3 GENERAL REMARKS ON C, H & N ANALYSERS

The brief descriptions above indicate the extent to which these complex and expensive instruments have simplified and speeded up the determination of carbon, hydrogen and nitrogen. There are many laboratories which are very satisfied with these instruments. They have filled a long-felt need and have transformed the role of the microanalyst, who can now deal with a very high demand if necessary.

Of course, the cost must always be considered, but the saving of the labour cost of assistants using the slower methods must be taken into account.

What does the operator do if the instrument becomes faulty? If, say, the normal practice is 30–40 determinations a day and this suddenly stops there will be a serious back-log of work, even though the instruments and the manufacturers have a good record for repair and maintenance. In the event of a malfunction the obvious thing is to call in the manufacturer. Would-be purchasers in more remote parts of the world need to consider very seriously, however, how they would be placed if in need of after-sales service.

This, then, is the prospective purchaser's dilemma. If one of these instruments is thought to be needed, then the need is really for two, so that each can be regarded as a reserve in case of breakdown of the other.

Otherwise, and preferably in any case, it is almost essential that at least one person in the laboratory must thoroughly understand the workings of the instrument so that essential spares can be kept in stock and repairs and replacements can be made. Of course, the skill of experienced operators in suspecting and noticing faults is invaluable. An automatic analyser is not like a washing machine or refrigerator which can be switched on and forgotten; it must be used by people who understand it.

An inexperienced person wishing to start up in micro-analysis at the present time would probably find that an automatic analyser would be the wisest choice, provided that there is sufficient work for the instrument, because so much has to be learnt by experience both with the traditional and the fully automatic equipment, that the user may as well begin to learn about the latter right from the start. However the automatic analysers are best suited to analysis of compounds of similar composition, and are thus ideal for such applications as quality control of a product, but if successive samples differ very widely in composition, the results may be unsatisfactory.

6.4 OTHER ANALYSERS

Kozłowski and Biziuk [2] have described a system for the simultaneous determination of hydrogen, nitrogen and halogen (other than fluorine), which can be automated for hydrogen and nitrogen determination. Compounds containing two halogens can also be analysed for both of them. The method can be used in conjunction with methods for simultaneous determination of carbon, hydrogen and halogen [3] or carbon, hydrogen and sulphur [4] to give the C, H, N, S, halogen ratios without the need to weigh the sample. Determination of atomic ratios without weighing the sample [2,4–7] is a comparatively little used technique, but is useful if not much sample is available. For this purpose Rezl [6,7] uses his own design of automatic C, H, N analyser [8]. A C, H, S analyser has been described [9].

The Technicon and Hewlett-Packard instruments are both designed for use of small samples (300-800 μg). The Technicon instrument uses a silica gel tube to remove water before the measurements of carbon dioxide, and then the water is released by heating and measured. The apparatus requires a rigid time schedule to be kept. The Hewlett-Packard instrument uses a gas chromatography system with an integrator and print-out system available.

A C, H, N analyser based on the Technicon design, incorporating automatic sample introduction, has been developed by Fraisse *et al.* [10].

The Perkin-Elmer 240 instrument has been adapted for use with an automatic sample-loader, and computerized operation and data-handling [11].

REFERENCES

[1] R. Belcher (ed.), *Instrumental Organic Elemental Analysis*, Academic Press, London, 1977.
[2] E. Kozłowski and M. Biziuk, *Mikrochim. Acta*, 1979 **I**, 1.
[3] E. Sienkowska-Zyskowska, *Dissertation*, Gdańsk Polytechnic, 1975.
[4] B. Kobylińska-Mazurek and E. Kozłowski, *Mikrochim. Acta*, 1978 **I**, 137.
[5] E. Kozłowskei and M. Biziuk, *4th Polish Conf. Anal. Chem. 1974*, Warsaw, 1974, p. 188.
[6] V. Rezl, *Mikrochim. Acta*, 1978 **I**, 493.

[7] V. Rezl, J. Uhdeová and J. Bursa, *Mikrochim. Acta,* 1979 **II**, 343.
[8] V. Rezl and B. Kaplamová, *Mikrochim. Acta,* 1975 **I**, 493.
[9] B. Kobylińska-Mazurek and E. Kozłowski, *Mikrochim. Acta,* 1978 **I**, 137.
[10] D. Fraisse, B. Cousin and C. Muller, *Mikrochim. Acta,* 1978 **II**, 9.
[11] W. R. Bramstedt and D. E. Harrington, *Microchem. J.,* 1979, **24**, 158.

CHAPTER 7

Oxygen

The feature that distinguishes the determination of oxygen in organic substances from that of most other elements is that it cannot be done by combustion of the sample in an excess of oxygen (unless the total quantity of oxygen present is very precisely known). Probably for this reason it has always been the most difficult of the common elements to determine, and it was not until 1939, when Schütze [1] established a semimicro method, later adapted to the micro scale by Zimmermann [2] and modified in 1940 by Unterzaucher [3], that the determination became fairly widely performed. The principle of the method is to heat a few mg of sample in an oxygen-free stream of an inert gas such as nitrogen, so that all the oxygen contained in the sample is converted into gaseous products (H_2O, CO, CO_2, O_2 etc.) which are converted into carbon monoxide by passage over granular carbon at about 1120°C. The carbon monoxide is then measured.

Unterzaucher passed the carbon monoxide over a heated hydrate of iodine pentoxide, to oxidize it to carbon dioxide, with liberation of an equivalent amount of iodine. This was titrated with sodium thiosulphate solution (after conversion into iodate as in the Leipert method, p. 97, to give a six-fold amplification). In other methods the carbon dioxide produced thus or by other means from the carbon monoxide was collected and weighed or titrated, etc.

The method has always caused difficulties and been subject to error. The temperature of the carbon is very close to that at which it begins to react with the silica tube, especially if the tube or the carbon is not pure. The very best fused quartz had to be selected, and tubes had to be washed in 40% hydrofluoric acid and perhaps worked near fusion point in a blow-pipe flame to produce a resistant surface. The carbon had to have a low ash content (less than 0.01%) and the presence of iron was known to cause trouble. It was recommended to wash the carbon for some hours with dilute hydrochloric acid. The carbon had to be amorphous and it was sometimes prepared as lamp black from benzene or acetylene, although various commercial sources were recommended.

A furnace to heat the carbon to 1120°C was not easy to design (although the constancy of its temperature was probably more important than its actual temperature) so that when Oita and Conway in 1954 found that 50% platinized carbon at 900°C could be used instead of carbon at 1120°C it appeared that some difficulties had been removed. However, in spite of claims to the contrary, many workers found difficulty with sulphur-containing samples (some used a cold trap to remove carbonyl sulphide or carbon disulphide).

The method really proved worthwhile only in laboratories large enough to have a sufficiently big demand for oxygen determination for the apparatus to be kept in good condition by continuous use. Even then, various problems may arise [4].

The calculation of the oxygen content from the volume of thiosulphate required is not often clearly explained; it is readily seen from the equations.

$$\tfrac{1}{2}O_2 + C \longrightarrow CO \qquad (1)$$
$$5CO + I_2O_5 \longrightarrow I_2 + 5CO_2 \qquad (2)$$

$$I_2 \xrightarrow{Br_2} 2IO_3^- \qquad (3)$$

$$IO_3^- + 5I^- + 6H^+ \longrightarrow 3I_2 + 3H_2O \qquad (4)$$
$$I_2 + 2S_2O_3^{2-} \longrightarrow 2I^- + S_4O_6^{2-} \qquad (5)$$

From (5) 1 ml of 1*M* thiosulphate ≡ 127 mg of iodine
From (4) 127 mg of iodine ≡ 175/6 mg of iodate
From (3) 175/6 mg of iodate ≡ 127/6 mg of iodine
From (2) 127/6 mg of iodine ≡ 70/6 mg of carbon monoxide
From (1) 70/6 mg of carbon monoxide ≡ 40/6 mg of oxygen
= 6.667 mg.

Because of the unsatisfactory nature of the classical method for oxygen, the author recommends that one of the following courses be taken. If there is sufficient demand for oxygen determination, a modern automatic elemental analyser such as the Perkin-Elmer Model 240 or the Carlo Erba Model 1106 should be bought and used, but if oxygen determinations are not often required the samples should be sent to a commercial laboratory specializing in this work. The modern analysers have made it easier for these laboratories to offer an efficient service.

A few notes on the most used automatic analysers are given below. Very useful descriptions and discussions of the most popular automatic analysers used for determination of oxygen are given in Belcher's monograph [5]. The Balzers Exhalograph EAO-202 can also be used [6].

7.1 THE PERKIN-ELMER MODEL 240

The conversion of this instrument from C, H & N determination to oxygen determination can be done in less than an hour, and useful results obtained in an hour or two, provided tubes such as the platinized-carbon reaction tube and other special tubes are prepared in advance. However, where circumstances justify it, it would be preferable to keep the apparatus permanently set up for oxygen determination.

The conversion for oxygen determination was described in a short paper by Culmo [7].

The C, H & N combustion tube is replaced by a similar quartz tube (the pyrolysis tube) packed with granular platinized carbon, 60–100 mesh copper, platinum gauze and silver gauze. It is heated to 975°C. The reduction tube is replaced by a similar tube (the oxidation tube) approximately one-third filled with wire form cupric oxide with 60–100 mesh copper on either side. This tube also contains some silver gauze and a little quartz wool to divide the zones. It is heated to 670°C.

Between the pyrolysis tube and the oxidation tube is fitted a U-tube containing 'Colorcarb' (sodium hydroxide on asbestos) to remove all acid gases emerging with the carbon monoxide from the pyrolysis tube. This tube and its fittings are supplied by Perkin-Elmer, but it is now thought preferable to use the improved design recommended for use on the same instrument for sulphur determination. The source of oxygen used in the C, H & N determination must, of course, be disconnected. The conversion can be done quite rapidly and furnace temperatures need be only slightly reduced. Detailed instructions on the conversion and operation of the Model 240 are provided by Perkin-Elmer with the conversion kit.

A sample of 1–3 mg of an organic substance is used for the determination. The carbon monoxide produced in the pyrolysis tube is oxidized to carbon dioxide in the oxidation tube. The carbon dioxide is measured in the same way as for the C, H & N determination (p. 72).

Signals are also recorded for hydrogen and nitrogen, but these are not likely to be reliable because the pyrolysis may produce compounds such as HCN.

Some workers have found that sulphur compounds give poor results although the apparatus recovers after a time. Fluorine compounds cannot be expected to give good results because they usually produce oxygen by attacking the quartz tube. Samples containing boron or metals may give difficulty, especially if the metal has an oxide stable at 900°C.

7.2 THE CARLO ERBA CHNO ANALYSER

The oxygen determination is done with this instrument in a permanent channel which is an alternative to the C, H & N channel. About 0.8–1.0 mg of sample is weighed in a silver capsule and fed automatically from a 24-compartment magazine into a pyrolysis zone at 1000°C, in helium. The oxygen in the gaseous pyrolysis products is converted into carbon monoxide over a special carbon-based formulation (containing nickel and platinum) which is effective at 1000°C. The carbon monoxide peak obtained chromatographically is integrated automatically and the figure related to that from known oxygen compounds. Peaks are also obtained corresponding to hydrogen and nitrogen gases from the sample, if any.

The oxygen channel contains platinum wool, ordinary activated charcoal, special activated charcoal containing nickel and platinum, quartz chips and its own chromatographic column containing molecular sieve 5 A at 120°C.

In contrast to the Perkin–Elmer apparatus, no difficulties have been reported with sulphur compounds, but chlorine-containing compounds give unsatisfactory results unless the method of Kirsten [8] is adopted. In this, the incoming helium stream is bubbled through a volatile chlorinated hydrocarbon.

With the Carlo Erba apparatus the weighing of volatile liquid samples both for C, H & N and for oxygen has given some trouble. Glass microcapillaries can now be purchased for gas chromatographic work, which have standard bore (1 μl ≡ 2.56 mm) so that correct amounts of sample can be more easily arrived at. The sealing of the liquid in the tube has also been made more satisfactory by placing the tube in a 1 mm diameter hole in a metal block, which may be cooled during sealing. The sealed capillary, about 6 mm in length, may then go into the 24-hole magazine. The capillary will burst as soon as it enters the pyrolysis chamber.

REFERENCES

[1] M. Schütze, *Z. Anal. Chem.*, 1939, **118**, 241.
[2] W. Zimmermann, *Z. Anal. Chem.*, 1939, **118**, 258.
[3] J. Unterzaucher, *Chem. Ber.*, 1940, **73**, 391.
[4] R. Belcher, G. Ingram and J. R. Majer, *Talanta*, 1969, **16**, 881.
[5] R. Belcher (ed.), *Instrumental Organic Elemental Analysis*, Academic Press, London, 1977.
[6] F. Ehrenberger, *Mikrochim. Acta*, 1977 **II**, 39.
[7] R. Culmo, *Mikrochim. Acta*, 1968, 811.
[8] W. Kirsten, *Microchem. J.*, 1977, **22**, 60.

CHAPTER 8

Chlorine

8.1 GENERAL DISCUSSION

Sufficient sample to contain about 1.7 mg of chlorine is burnt in an oxygen flask (Section 3.2) containing a little sodium hydroxide solution and hydrogen peroxide. The acidity of the resulting solution is adjusted, 100 ml of almost anhydrous propan-2-ol are added, followed by diphenylcarbazone indicator, and the chloride is titrated with 0.005*M* mercuric nitrate solution to the first appearance of a violet colour. Quantities of water used are limited throughout so that the final solution is almost non-aqueous and mercuric chloride is not ionized. The end-point is very satisfactory and the whole procedure simple. Bromine can be determined in the same way but the method described in Chapter 9 is better. This method for the determination of chlorine gives consistently good results and is very simple, especially as the end-point of the titration is very easy to see. However, highly chlorinated hydrocarbons such as hexachlorobenzene may be difficult to combust; this problem may often be overcome by adding a few mg of sucrose to the sample.

Many of the other methods that have been used now seem inferior, some needing more elaborate equipment and others being slower. However, the author considers that the Carius method, completed by weighing silver halide, can still prove valuable in certain cases. Potentiometric titrations of the halides, though capable of automation, give more difficulty with bromine determinations, probably because the oxy-acids of bromine are more readily formed than those of chlorine, and also the electrodes used, such as those of silver metal, need frequent cleaning and regeneration. In addition, although it is possible to titrate chloride and bromide separately when present together, the titration volumes are not quite stoichiometric, probably because of occlusion of silver chloride in the bromide precipitate. For similar reasons an attempt to use the method of Vieböck [1] for routine rapid determination of chlorine and bromine is not sufficiently reliable even when the mercuric oxycyanide solution is purified by dialysis and the matching method is used. It has been suggested [2] that the method is too sensitive to temperature, volumes and quantities although the matching method is meant to compensate for such errors. In any case the method is more troublesome than that recommended in this chapter. The mercuric nitrate titration described here includes the improvements suggested by Cheng [3] and White [4], the most important of which seem to be control of the acidity and the addition of sufficient alcohol, (propan-2-ol or ethanol) for the titration solution to contain about 1 part of water to 4 parts of alcohol. Recovery of most of the more expensive propan-2-ol is easy, and desirable because it gives a consistent halogen-free product.

Although the titration end-point is very sharp, the reaction may not be exactly stoichiometric. For this reason the titrant should be standardized against the combustion products of pure standard substances, and the procedure should be as consistent as possible. This also tends to eliminate the effect of any error in the blank value.

8.2 METHOD

8.2.1 Apparatus

Combustion flasks. Iodine flasks provided with platinum baskets as described in Section. 3.2.

Burettes with reservoirs and automatic filling devices, 10 ml, graduated at 0.02 ml intervals for the mercuric nitrate and at 0.05 ml intervals for the nitric acid.

8.2.2 Reagents

Mercuric nitrate solution, 0.005*M*. Mercuric nitrate solid (A.R.) can be represented by the formula $Hg(NO_3)_2 . \frac{1}{2} H_2O$, formula weight 333.6. It may appear to be slightly moist but is at least 99.0% 'pure'. If this material is dissolved in water, basic mercuric nitrate separates, so the solution must be made up in dilute nitric acid and filtered before use. To make 4 litres dissolve 7.0 g of solid in 400 ml of 0.01*M* nitric acid. Dilute to 4 litres with 0.01*M* nitric acid. Allow to stand for 48 hr, filter and standardize against pure sodium chloride. Further dilution, if required, should again be with 0.01*M* nitric acid, and the solution should be finally standardized by the determination of chlorine in pure standard organic substances, of which the most commonly used is *p*-chlorobenzoic acid. Mercuric nitrate solution can also be made from basic mercuric nitrate [$2Hg(NO_3)_2.HgO$, formula weight 865.8], from mercuric oxide (HgO, F.W. 216.59) or from metallic mercury (Hg, atomic weight 200.6) by dissolution in nitric acid. The basic mercuric nitrate has been found to be quite satisfactory, although only the reagent grade is available, if it is made up in 0.01*M* nitric acid instead of 0.01*M*. Metallic mercury is best dissolved in 4*M* nitric acid.

Propan-2-ol. A blank should be determined on each batch. It is recommended that all titration solutions should be combined after use and the propan-2-ol recovered by distillation after most of the water has been removed by treatment with solid sodium hydroxide. This routine recovery results in the propan-2-ol being free from halides and sufficiently low in water content.

Take 1.5 litres of titration residues in a 3 litre flask and add about 160 g of sodium hydroxide flakes. Shake occasionally, with care (because of the caustic solution) and a lower aqueous layer will separate. Let stand for 24 hr, then transfer to a 2 litre separating funnel and discard the lower aqueous layer. Transfer the upper layer to a dry 3 litre conical flask, add about 80 g of sodium hydroxide flakes, shake well and let stand overnight. Decant from the solid into a 3 litre round-bottomed flask equipped for distillation. (Use the solid which remains for the first treatment of the next batch of titration residues, with addition of about 100 g more sodium hydroxide). Combine two 1.5 litre batches, add a few pellets of sodium hydroxide and some anti-bumping granules, and distil off the alcohol from a steam-bath through a 20 cm Dufton fractionating column, a still-head and a double-surface condenser, using a silica-gel guard-tube to exclude moisture. The alcohol boils at 82°C; adjust the steam so that the distillation rate is 1 or 2 drops/sec. This procedure, when made a regular practice, has been found to be very easy to carry out, and provides an alcohol of consistent quality and of sufficiently low water content (about 3%).

Diphenylcarbazone solution, 0.1% in ethanol. Since the solution is not very stable it is advisable to prepare not more than 50 ml at a time.

Bromophenol Blue, 0.1% solution in ethanol.

Barium nitrate, 0.1*M*. Dissolve 26 g of barium nitrate (A.R.) in 1 litre of pure water.

Sodium hydroxide, 0.1*M*. A commercial volumetric solution is convenient.

8.2.3 Procedure

After every determination, empty the contents of the combustion flask into the reservoir for propanol recovery. Rinse the flask thoroughly with pure water (filling it at least once, to remove all organic vapours) and shake it out, but leave it damp. Add the absorption solution – 2.5 ml of 0.1*M* sodium hydroxide, 0.5 ml of 100 vol hydrogen peroxide and 1.0 ml of water. Weigh a sample as described in Section 3.2 and roll it up in the filter paper. Take about 9 mg for 20% of Cl, 12 mg for 15%, 18 mg for 10% etc. Combust as described in Section 3.2. After 10 min, or longer, remove the stopper and rinse it and the platinum into the flask with exactly 10 ml of water. This is done either with the wash-bottle described by Fill and Stock [5] or by adding 10 ml of water to a small fully discharged wash-bottle and fully discharging it again, using the jet for rinsing. Keep the stopper and platinum uncontaminated for a subsequent rinse with the propanol. Add 2 drops of Bromophenol Blue solution and add 0.1*M* nitric acid from the burette until the blue (alkaline) solution turns just yellow (acid). Add 1 ml of 0.1*M* barium nitrate. This is needed only for sulphur-containing substances but is added in all cases so that the blank is unaffected. Add 0.5 ml of 0.1*M* nitric acid from the burette. Measure out 100 ml of propanol, and use it to rinse thoroughly (into the flask) the stopper, the platinum, the collar and the walls of the flask. Retain about 10 ml of propanol for a subsequent rinse. Add 0.5 ml of diphenylcarbazone solution. Titrate with 0.005*M* mercuric nitrate (about 5 ml will be required) from the pale yellow to a faint violet tint at the end-point (v ml). The solution must be kept well agitated and the mercury solution added slowly; near the end-point add the remainder of the propanol, rinsing in the platinum and stopper and the flask walls. The approach of the end-point will be clearly heralded by a violet colour which disappears on further mixing. Over-titration produces a deepening of the violet colour, which is easily seen. Do a blank determination in the same way, but using a sample which is halogen-free (b ml). The percentage of chlorine in the sample is $f(v-b)/w$ where w = weight of sample (mg) and f is the chlorine equivalent (mg per 1 ml of the mercuric nitrate solution, determined by calibration against standard compounds).

Liquid samples may be too volatile to weigh in the open. If so:

(a) weigh the sample in a glass capillary and burn it as for a sulphur determination on a volatile liquid (Belcher and Ingram combustion), using the same absorption solution as for the flask combustion but rinsing with only 10 ml of water, followed by propanol;

or (b) do a Carius chlorine determination [here although the sample is volatile, enough of it (45 mg) can be taken in a narrow weighing tube for the loss during weighing to be negligible];

or (c) weigh in a gelatine capsule as described on p. 33. The blank must be redetermined.

The Carius method is usually the most convenient of these alternatives.

Bromine. If bromine is present with chlorine it will consume a corresponding volume of mercuric nitrate solution. This can be subtracted if the bromine is determined independently as described in Chapter 9. The bromine determination alone, however, may often be sufficient identification of the sample (Chapter 9).

The following is an example of the calculation (assuming that the mercuric nitrate is exactly 0.005*M*). A 16.367 mg sample required 12.73 ml of mercuric nitrate solution. The sample had been found to contain 21.0% of Br, i.e. $16.367 \times 0.210 = 3.44$ mg of Br. Each ml of the mercuric nitrate solution was equivalent to 0.799 mg of Br. Therefore 3.44 mg of Br was equivalent to $3.44/0.799 = 4.30$ ml of mercuric nitrate solution. Hence the remainder, i.e. $12.73 - 4.30 = 8.43$ ml, was equivalent to the chlorine present which was, therefore, $8.43 \times 35.45/16.367 = 18.26\%$ Cl. (If it were to be assumed, as was true in this example, that there were 2 atoms of chlorine present for each atom of bromine, then the titration volume for the bromine alone would have been $12.73/3 = 4.24$ ml). The sample theoretically contained 18.6% of Cl and 21.0% Br.

Iodine. When iodine is present, the hydrogen peroxide should be omitted because it would produce elemental iodine. Instead, use as the absorption solution 1 ml of fresh saturated sulphur dioxide solution and 4 ml of water. After combustion and standing, rinse the stopper and platinum with 10 ml of water, as usual, bring to the boil rapidly with shaking to remove sulphur dioxide and continue boiling for about 1 min, cool and proceed as for chlorine and bromine. The barium nitrate will remove the sulphate that is produced from the sulphur dioxide. 1 ml of 0.005*M* mercuric nitrate ≡ 1.269 mg of I.

If chlorine (or bromine) is present with iodine and the iodine is determined by Leipert's method, Chapter 10, then the chlorine (or bromine) content can be calculated as when chlorine is present with bromine.

REFERENCES

[1] F. Vieböck, *Chem. Ber.*, 1932, **B65**, 496.
[2] J. E. Fildes and A. M. G. Macdonald, *Anal. Chim. Acta*, 1961, **24**, 121.
[3] F. W. Cheng, *Microchem, J.*, 1959, **3**, 537.
[4] D. C. White, *Mikrochim. Acta*, 1961, 449.
[5] M. A. Fill and J. T. Stock, *Analyst*, 1944, **69**, 149.

CHAPTER 9

Bromine

9.1 GENERAL DISCUSSION

For the determination of bromine [1-5], a sample of about 10 mg of the organic substance is combusted in an oxygen flask and the products are absorbed in a solution of hypochlorite and a buffer. After standing, opening and rinsing in of the platinum sample holder, the solution is heated and to the boiling solution an excess of sodium formate solution is added. The solution is allowed to stand to complete the reduction of the hypochlorite, any chlorine-containing vapour is blown away and then the solution, which now contains the bromine as bromate, is cooled to room temperature. A large amount of sulphuric acid is added (to overcome the buffer), followed by 1 g of potassium iodide and a trace of ammonium molybdate as catalyst. Iodine is liberated according to the equation

$$HBrO_3 + 6HI = 3I_2 + 3H_2O + HBr$$

so 6 atoms of iodine are produced from 1 atom of bromine in the original sample. Therefore the iodine is in sufficient quantity to be titrated with 0.05*M* sodium thiosulphate. A blank should be determined and subtracted; it is usually about 0.05 ml.

It is advisable to be consistent in procedure, including the time that the solution spends on the hot-plate.

Many methods are available for the determination of chlorine and bromine but bromine often appears to be the more difficult to determine, probably because the oxy-acids of bromine are readily formed. The present method avoids the problem of converting these into bromide, and has the additional, often very useful, advantage of being specific for bromine (in absence of iodine).

9.2 METHOD

9.2.1 Apparatus

Conical Pyrex flasks (500 ml) with B24/29 ground-glass sockets.
Platinum baskets, as described in Section 3.2.

A 5 ml tilt gauge (automatic dispenser) for the phosphate.
A rapid 25 ml burette, with rubber tubing and pinch-clip (not a glass tap) for 1*M* sodium hydroxide.
A 5 ml rapid dispenser (filled by squeezing a polythene container) for the sodium hypochlorite solution.
A small electrical hot-plate, with 3 heat settings.
A 10 ml tilt gauge for 9*N* sulphuric acid.
A 10 ml burette, graduated to 0.02 ml, with reservoir (1 litre) and pressure-filling device for 0.05*M* sodium thiosulphate.

9.2.2 Reagents

Ammonium molybdate, A.R. 5% solution in water.
Sodium hypochlorite solution. About 1*N*, in 0.1*M* sodium hydroxide.
Sodium dihydrogen phosphate solution, 20%. Put 60 g of the solid in a 350 ml bottle which has the top edge of its label at a level corresponding to about 300 ml, add water, dissolve and dilute to top of label.
Sodium hydroxide solution, 1M. A commercial solution is convenient but not essential.
Sodium formate solution, 50%. Dissolve 150 g of the solid in about 200 ml of water and dilute to 300 ml.
Sulphuric acid, 9N. Add 126 ml of conc. sulphuric acid carefully to about 300 ml of water. Allow to cool and dilute to 500 ml.
*Sodium thiosulphate solution, 0.05*M. Prepare 6 litres from 3 ampoules of commercial concentrate (each intended to make 1 litre of 0.1*M* solution), adding about 6 g of borax as stabilizer before making up to volume. Alternatively dissolve 75 g of the solid pentahydrate and 6 g of borax in water and dilute to 6 litres. Use freshly boiled and cooled demineralized water. Keep the solution in the dark and standardize it regularly against weighed portions of potassium iodate. Before standardization mix the portion in the burette reservoir with the main stock of solution. In this way the change in strength is more gradual and predictable.
Iodine indicator solution, 40%. Dissolve 40 g of iodine indicator (May and Baker) in water and make up to 100 ml (mark on dropping bottle). Alternatively use 5% starch solution.

9.2.3 Procedure

The combustion procedure is the same as for chlorine, iodine, fluorine and sulphur determinations except for the absorption solution. Set the hot-plate so that it will boil 50 ml of water in a 500 ml conical flask in 4 min. Clean a 500 ml combustion flask thoroughly with hot water (iodide from a previous titration must be completely removed), filling it at least once to remove any organic vapours. Rinse it with demineralized water and shake out the surplus. Put into the flask 5 ml of sodium dihydrogen phosphate solution and 1.5 ml of 1*M* hydroxide. Take about 10 mg of sample (or that amount which will give a titration of 5–8 ml), as described in Section 3.2. Pass oxygen into the combustion flask rapidly for 30 sec, then slowly, while adding 5 ml of sodium hypochlorite solution, mixing the solutions and wetting all the inside of the flask with the mixture; then immediately combust the sample as described in Section 3.2, and leave for at least 5 min for the products to be absorbed. Remove the stopper, rinsing it, the joint and the walls of the flask thoroughly with pure water so that the total volume is about 50 ml (conveniently marked on the flask). Stand the flask on the centre of the hot-plate and boil the solution for 1 or 2 min (always for exactly the same total time, about 6 min, on the hot-plate). Add 7 ml of sodium formate solution and swirl the solution gently for 15 sec. Stand the solution on the bench for 4 min, blow away any chlorine vapour, rinse down the walls with about 30 ml of water and cool the solution under running water to 25°C or below. Add 10 ml of 9*N* sulphuric acid and about 1 g of potassium iodide, gently swirling to dissolve it, and 2 drops of ammonium molybdate solution. Allow 2 min for all the iodine to be formed, then titrate with 0.05*M* thiosulphate, adding about 1 ml of indicator near the end-point. Subtract the mean titration volume for a blank determined in the same way, several times, by combustion of a bromine-free standard.

Calculate the bromine content from the formula $\%\ \mathrm{Br} = 13.32\,(v-b)F/w$ where v = volume of thiosulphate used for the sample (ml), b = volume used for the blank (ml), w = weight of sample taken (mg), F = molarity of the thiosulphate solution.

Recovery of bromine has been found to vary from 97 to 100% but is generally above 99%.

Volatile liquids (and solids). When the sample is too volatile to be weighed in an open vessel before transfer to the filter paper or even loses weight when in the paper roll, it can usually be analysed by the Carius method (Chapter 11). The loss in weight is then often negligible for about 20 mg of sample taken in a narrow weighing tube, because only a small surface area is exposed. If the laboratory has the Carius equipment readily available this is the easiest solution to the problem of volatile samples, but silver chloride, bromide and iodide will be collected and weighed indiscriminately, if present, so the method is not specific.

Another possibility is to weigh the sample in a sealed glass capillary and to use the Belcher and Ingram apparatus (p. 114) with an alkaline absorption solution, the phosphate buffer and sodium hypochlorite being added later. Calibration with standard samples is essential.

The sample may also be weighed in a small sealed plastic pouch or capillary as described in Section 3.2, and burned in the oxygen flask, but it may be difficult to find plastic giving a reproducible blank.

REFERENCES

[1] J. F Alicino, A. Crickenburger and B. Reynolds, *Anal. Chem.*, 1949, **21**, 755.
[2] W. Schöniger, *Mikrochim. Acta,* 1956, 869.
[3] G. Ingram, *Methods of Organic Elemental Microanalysis,* Chapman & Hall, London, 1962, p.205.
[4] R. Belcher, *Submicro Methods of Organic Analysis,* Elsevier, Amsterdam, 1966.
[5] R. Reverchon, *Bull. Soc. Chim. France,* 1972, 831.

CHAPTER 10

Iodine

10.1 GENERAL DISCUSSION

A sample of 10-20 mg (according to iodine content) of the organic substance is combusted in an oxygen flask and the products are absorbed in an alkaline solution. After standing, opening and rinsing of the flask, an excess of bromine in glacial acetic acid is added, to convert iodine into iodate.

$$I_2 + 5Br_2 + 6H_2O = 2HIO_3 + 10HBr$$

Formic acid is then added to destroy the excess of bromine.

$$HCOOH + Br_2 = 2HBr + CO_2$$

Complete removal of Br_2 is confirmed by the addition of a drop of Methyl Red solution, which will not be bleached if no bromine is left. Dilute sulphuric acid and an excess of potassium iodide are added. Iodine is now liberated according to the equation

$$IO_3^- + 5I^- + 6H^+ = 3I_2 + 3H_2O$$

so 6 atoms of iodine are produced from the one atom of iodine in the original sample. The iodine produced is sufficient to be titrated with 0.05*M* sodium thiosulphate solution; a blank correction is not normally required.

This amplification method, which was modified by Leipert [1] for use on the micro-scale, has long been found to be highly satisfactory, but it became even more so when combined with the oxygen-flask combustion of the sample [2-4].

Here, sodium thiosulphate of concentration 0.05*M* is recommended because it is more stable than more dilute solutions. Sufficient sample should be taken to give a titration of 5 ml or more. This is rarely found to be difficult.

10.2 METHOD

10.2.1 Apparatus

Combustion flasks and platinum baskets, as described in Section 3.2.
Burette, 10 ml for 0.05*M* sodium thiosulphate, as used for bromine determination (p. 94).

10.2.2 Reagents

Bromine solution. A solution of bromine and sodium acetate in acetic acid. Into a 2 litre flask weigh 200 g of hydrated sodium acetate, add 100 ml of distilled water and heat gently on a steam-bath until dissolved. Dilute to 2 litres with glacial acetic acid. Take 500 ml of this stock solution, and add to it 5 ml of bromine (*safety pipette*).
Absorption solution. An approximately 0.3*M* sodium hydroxide solution. Dilute 150 ml of 1*M* sodium hydroxide to 500 ml in a polythene bottle.
Formic acid 90% w/w
Methyl Red solution, 0.25% in ethanol.
Potassium iodide
Sulphuric acid 1*M.* Laboratory bench reagent.
Iodine indicator solution. Prepare as on p. 94.
Sodium thiosulphate, 0.05*M*. Prepare as described on p. 94.

10.2.3 Procedure

The flask combustion procedure is described in Section 3.2.

Clean the 500 ml combustion flask thoroughly with hot water (iodide from a previous titration must be completely removed), filling it at least twice to remove organic vapours. Rinse with pure water and shake to remove excess but leave it moist. Add 20 mg of absorption solution to the combustion flask. Take enough sample to give a 7-8 ml titration, using the following figures for guidance: for 70% of I take 11 mg, for 40% or less take 20 mg. Carry out the combustion as described in Section 3.2. The presence of iodine will usually be indicated by the appearance of iodine vapour. When combustion is complete, wait until the flask is cold, then shake it well, keeping it in the inverted position. Return it to its normal position, moistening the joint at the same time. Leave stoppered for at least 5 min (normally 10 min or more). Remove the stopper, rinsing it in thoroughly with pure water. Also rinse down the joint and walls of the flask. Add from a measuring cylinder 12 ml of bromine solution down the sides of the flask and swirl well. Leave for 2-3 min away from the vicinity of the titration area, preferably in the fume cupboard where the bromine solution is stored. Add from a dropper about 1 ml of formic acid down the side of the flask. Swirl well and leave for 1-2 min before blowing vapour from the inside of the flask (compressed air may be used in moderation for this). Add 1 drop of Methyl Red solution. If the Methyl Red is bleached, bromine is still present. Continue adding formic acid and Methyl Red in drops, with pauses, until the colour of Methyl Red persists. Add about 1 g of potassium iodide and a measure (4–5 ml) of 1*M* sulphuric acid. Titrate the liberated iodine with 0.05*M* sodium thiosulphate until the solution is pale yellow, then add 6 drops of iodine indicator. Continue titrating dropwise until the blue colour of the starch-iodine complex is just removed. As the iodine titrated is equivalent to 6 times the iodine present in the sample, 1 ml of 1*M* thiosulphate ≡ 126.9/6 mg of iodine in the sample.

The % iodine in the sample = $21.15\,(V-b)F/w$ % where V ml is the sample titration volume, b ml is the blank (normally zero for iodine determination), w mg is the weight of sample taken and F is the molarity of the thiosulphate solution.

Use *o*-iodobenzoic acid (I = 51.17%) as standard (keep it over P_2O_5 in a desiccator).

To standardize the 0.05*M* thiosulphate solution weigh approximately 10 mg of dry potassium iodate (A.R.) into a clean 250 ml flask, add about 25 ml of water and dissolve. Proceed as above, starting at 'Add about 1 g of potassium iodide'.

REFERENCES

[1] T. Leipert, *Mikrochim. Acta,* 1938, 73.
[2] G. Ingram, *Methods of Organic Elemental Microanalysis,* Chapman & Hall, London, 1962, pp. 179, 206.
[3] W. Schöniger, *Mikrochim. Acta,* 1955, 123.
[4] W. Schöniger, *Mikrochim. Acta,* 1956, 869.

CHAPTER 11

Halogens by the Carius method

11.1 DISCUSSION OF THE METHOD

In this method [1-3], the sample (not less than 20 mg if sufficient is available) is decomposed by heating at 280-300°C with 0.3 ml of fuming nitric acid and about 100 mg of silver nitrate in a sealed glass tube. The halogen in the sample is converted into the corresponding insoluble silver halide, which is washed into a beaker and heated to assist coagulation. This precipitate is transferred quantitatively to a pre-weighed sintered-glass crucible, washed, dried at 150°C, and weighed. (An abbreviated method may be used for ionizable halogens: the sample is dissolved in water and silver nitrate solution added, and the silver halide precipitate formed is determined gravimetrically in the same manner.) More than 20 mg of sample may be taken when the halogen content is low but 45 mg of sample should not be exceeded even with a full length micro-Carius tube, otherwise it will be likely to explode whilst in the heater. When halogen is present as an accidental impurity at about the 1% level this method can give quite precise and useful results. A chlorine content as low as 0.1% can be easily seen and roughly determined. (This chapter does not of course, refer to the determination of fluorine since silver fluoride is soluble in water).

The method described has proved invaluable over many years but recently its principal use has been in the determination of chlorine, bromine or iodine in fairly volatile liquids which are not easy to weigh for oxygen-flask combustion but which can often be weighed and sealed in a Carius tube with satisfactory precision. Liquid samples are usually sufficiently plentiful to permit taking enough to minimize the relative losses on weighing a volatile liquid in a narrow cup.

The method described uses ordinary small sintered glass crucibles, which are weighed against a similar one as counterpoise (as suggested by Clark [4]). The contents of the Carius tube are transferred to a beaker and then to the filter crucible. The more traditional micro-method is to dilute the contents of the Carius tube, heat the mixture for some time in a steam-bath, and then to use a narrow siphon tube and suction to transfer the precipitate to a weighed filter tube. Although this traditional procedure gives better precision than the one described here, it demands much more skill. The modified method gives sufficient precision when carried out with care and some experience. When it is possible to take 20 mg of sample and to obtain at least 20 mg of silver halide the precision obtained is such that the results should be within 0.1% of the true value. The method is so reliable that it is not necessary to carry out duplicate determinations. It is of instructional value, because the operator can see the chemical changes, from sample to silver halide, in such an obvious manner. The method is not impracticably slow: if nine tubes are put on to heat overnight, all nine determinations can be completed by one person in a working day.

If it is required to determine ionizable chlorine, for example the chlorine in the hydrochloride of a bromine compound, the sample can be weighed into a 30 ml beaker and dissolved in water. Nitric acid is added, the solution is heated, excess of silver nitrate is added and the precipitate is coagulated by heating. The determination is completed by collecting the precipitate on a weighed sintered-glass crucible, provided there is no interference by the organic part of the molecule.

If the sample contains more than one halogen, for example, chlorine and iodine, a mixture of silver chloride and silver iodide will be obtained. If, however, the iodine is determined by a Carius method in which it is oxidized to iodic acid as described by White and Secor [5], separate determination of chlorine and iodine is possible. Instead of silver nitrate, mercuric nitrate is put in the dry Carius tube, and after the heating, nitric and nitrous acids are removed with urea and sulphamic acids. The iodine is oxidized with bromine and determined as described in Chapter 10.

11.2 METHOD

11.2.1 Apparatus

Beakers, 30 ml. A set of 6 borosilicate glass beakers, 30 ml, wide form, labelled A–F, kept in a box ready for use.

Pyrex stirring rods. About 8 cm long, 2.5 mm diameter with flame-polished ends. These are kept with the beakers.

Glass rod, 2.5 mm diameter, with rounded end, for loosening silver halide from the inside of the Carius tube.

Beakers, 400 ml. Used to cover the 30 ml beakers.

Sintered-glass crucibles. Capacity 8 ml, plate diameter 15 mm, height above plate 35 mm, approximate weight 8 g, porosity 4 (5–10 μm). A set of 6 crucibles and one counterpoise crucible is kept in readiness in a desiccator.

Electric oven. Set at 150°C approximately.

Desiccator. Scheibler pattern, 8 or 10 in. diameter, fitted with a circular sheet of aluminium drilled with holes large enough to accommodate the bottoms of the sintered-glass crucibles.

Forceps. For handling crucibles etc.
Chamois leather. Impregnated with 1% glycerol solution, then left to dry at room temperature. Used to wipe the exterior of the crucible just before drying in the oven.
Tube rack. Test-tube rack with 12 holes, each numbered, to accommodate the Carius tubes when samples are being weighed (used for all Carius work).
Cleaning wire. Stainless-steel wire 16 s.w.g., with pipe cleaner wrapped spirally round end and giving a tight fit, used to clean new Carius tubes.

Other apparatus is as described in Section 3.1.

11.2.2 Procedure

Put about 100 mg of silver nitrate into a dry Carius tube. Tap it well down so that none adheres where the glass is to be heated for sealing. (After practice the correct amount of silver nitrate can be judged without weighing. It roughly fills the hemispherical end of the tube. It is essential that there should be enough to react with the halogen in the sample and give an excess).

Measure 0.3 ml of fuming nitric acid into the Carius tube from a safety pipette.

Weigh the sample in the weighing tube, (p. 28) using the glass stand and a glass counterpoise. When there is enough sample available, take 20 mg if the halogen content is expected to be about 25% Cl, 43% Br or 54% I, and more or less in proportion to yield about 20 mg of silver halide. If less sample is available the precision is not so high. Drop the cup into the Carius tube which contains the silver nitrate and nitric acid and seal it immediately as described in Section 3.1.

Heat the sealed tubes in the Carius heater for 3½ hr. Release the pressure in the cold tube by applying a flame to the tip of the capillary until the gases escape through the softened glass. Then cut the tube open at the shoulder, using the blow-pipe flame in order to avoid producing glass chips which might be collected with the silver halide. To do this take a glass rod or piece of used Carius tube and seal it on to the end of the capillary of the Carius tube in such a way as to form a handle and also to make the tube gas-tight again. Then, rotating the tube all the time, heat a ring round the shoulder of the Carius tube. The heating will cause an increase in the pressure in the tube and as the glass softens it will expand and eventually blow a hole. The edges of the hole should be immediately attacked with the oxygen-gas flame until the tube is left open with a flanged mouth which can be improved with a metal reamer while the glass is still soft. The flange lips should be thickened a little in the flame and the mouth should be large enough to permit the weighing cup to be taken out.

When the Carius tube is cold add about 1 ml of water to it. Shake it gently, because nitrous fumes are usually produced, and invert the tube with its mouth inside a 30 ml beaker. Gentle tapping will usually cause the weighing tube to leave the Carius tube and fall into the beaker with a bead of silver nitrate attached to it, containing most of the silver halide produced during the heating. (When hot the silver nitrate is liquid and it solidifies in a bead on cooling.)

Next, carefully remove the small amount of silver halide usually left in the Carius tube and transfer it with water into the beaker. Use small rinses and gently stroke the inside of the tube with a rounded-end glass rod to remove any silver halide clinging to the inside of the tube. It has been found that an old-fashioned glass wash-bottle, with a jet that is not too large or too small, used skilfully, is best for removing the last traces of solid silver halide from the Carius tube (by 'hydraulic mining' with a jet of water directed at the bottom of the upturned tube).

Rinse the weighing tube with distilled water into the beaker, stroking it with the small glass rod, which is then left in the beaker, and rinse with water. (After repeated use the weighing tube progressively acquires a brown colour due, no doubt, to the silver nitrate, but it can be used repeatedly if rinsed with water, dried in the oven and cooled in or near the balance for several hours before re-use).

Heat the 30 ml beaker on the steam-bath. The precipitate of silver halide must coagulate and at the same time all the excess of silver nitrate must go into solution. Silver chloride, silver bromide and silver iodide have characteristic appearances which should be observed (in colour, white, pale yellow and stronger yellow respectively), but in daylight the silver chloride and bromide rapidly turn a bluish colour (but the weights are not affected). Silver iodide is the least soluble in pure water but it dissolves in concentrated silver nitrate solution. Therefore, it usually separates in a finely divided form and must be left on the steam-bath for long enough to coagulate and produce a clear supernatant solution. Chloride and bromide precipitates should be left on the steam-bath for at least 15 min and the iodide will probably need longer.

Place a dried, cooled and weighed filter crucible in an adaptor in the filter flask and apply gentle suction. Decant the supernatant solution through the filter crucible, wash the silver halide once in the beaker with water, then transfer the precipitate to the crucible with the aid of the wash-bottle jet. The precipitate receives adequate washing with water in this process. After the water has been sucked through, wash the filter thoroughly with ethanol, dispensed through a filter paper and funnel.

Dry the silver halide and crucible in the oven at 150°C for 15 min. Dry the counterpoise crucible in the same way. Cool the crucibles in the desiccator for 1 hr before weighing.

Silver chloride contains 24.74% chlorine, silver bromide 42.55% bromine and silver iodide 54.05% iodine. Calculate the halogen content of the sample.

If it is not known which halogens are present, the result can be expressed as % AgX; that is to say the number of parts of silver halide derived from one hundred parts of sample. The result may be used to verify or otherwise a postulated formula.

If bromine or iodine has been determined separately by specific methods then the chlorine content can be calculated from the % AgX.

REFERENCES

[1] J. Grant, *Pregl, Quantitative Organic Microanalysis,* Churchill, London, 1951, p. 118.

[2] A. Steyermark, *Quantitative Organic Microanalysis,* 2nd Ed., Academic Press, New York, 1961, p. 316.

[3] J. B. Niederl and V. Niederl, *Micromethods of Quantitative Organic Elementary Analysis,* New York, 1938, Chapter VI.

[4] E. P. Clark, *Semimicro Quantitative Organic Analysis,* Academic Press, New York, 1943, pp. 54–57.

[5] L. M. White and G. E. Secor, *Anal. Chem.*, 1950, **22**, 1047.

CHAPTER 12

Fluorine

12.1 DISCUSSION OF THE METHOD

Organic substances containing fluorine are combusted in oxygen flasks in the usual way, except that quartz flasks are used to avoid low results which might be caused by fluorine reacting with the boron present in borosilicate glass. Oxygen-flask combustion has been found generally satisfactory except for very highly fluorinated compounds, especially perfluoro compounds (organic substances in which all hydrogen atoms have been replaced by fluorine atoms). For such compounds fusion with potassium metal in a closed nickel bomb is necessary.

After an oxygen-flask combustion, the fluoride ions are determined colorimetrically with alizarin fluorine blue [1,2]. No more than 12 μg of fluorine are needed in the final 100 ml of solution to give satisfactory precision.

The method described follows closely that described by Johnson and Leonard [1]. In the oxygen-flask combustion nearly all substances containing fluorine will give a quantitative yield of fluoride, with just water as absorption solution and with no additional oxidizing agents to assist the combustion or decomposition. Some highly fluorinated substances, especially perfluoro compounds, (e.g. PTFE) cannot be expected to decompose in this way. They have to be heated with potassium metal in a sealed metal bomb such as the Parr bomb or the bomb described by Belcher and Tatlow [3] and also by Ingram [4].

When alizarin fluorine blue (AFB) is used, quite small samples are sufficient, provided that a suitable balance is available, and even less is required when the fluorine content is high. This makes it easier to achieve complete decomposition. The amount of fluoride actually required in 100 ml of solution is 9–15 μg, when a 4 cm spectrophotometer cell is used.

The method described here incorporates several improvements on the original procedure of Johnson and Leonard. Lanthanum has been found to give a more intense colour than cerium, succinate has replaced acetate as buffer, and 20% acetone in the final solution has also been found to enhance the colour [5–8]. Hanocq and Molle [9] found that dimethylsulphoxide was even more effective and that propan-2-ol was only slightly less effective than acetone, and that it gave a more stable solution in the spectrophotometer cells.

Ion-selective fluoride electrodes have been available for some time now. A procedure based on use of the fluoride electrode, for example that of Shearer and Morris [10] should be the method of choice when many fluorine determinations have to be made. However, in many laboratories the number of fluorine determinations required will not justify the expense of acquiring an electrode and a suitable potentiometer, so one of the more tedious chemical methods will have to be used.

In the past, fluorine was notoriously difficult to determine. Substances were often decomposed in an oxygen-hydrogen flame [11]. Fluoride ions were titrated with thorium nitrate, although the titration presented many difficulties overcome only by strictly consistent procedures or the use of a photometric titration as described by Steyermark [12]. Fluoride has also been determined gravimetrically by precipitation as lead chlorofluoride. Disadvantages were the non-stoichiometry of the precipitate and the need for comparatively large amounts of fluorine (10–20 mg).

The AFB method given here is, if anything, too sensitive. Determinations can easily be carried out on 1 mg of sample if a suitably sensitive balance is available: otherwise considerable dilution will be necessary. If interfering elements (e.g. phosphorus) are present, it may be necessary to separate hydrofluoric acid by Willard and Winter distillation [13]. Fortunately this is rarely necessary. Most present-day 'commercial' fluorine compounds and intermediates usually contain either $-CF_3$ groups or have fluorine substitution in benzene or analogous rings. The flask combustion method hardly ever fails for such compounds.

An automated method based on hydrothermal decomposition and photometric titration with thorium has recently been described [14].

12.2 METHOD

12.2.1 Apparatus

Combustion flasks, 500 ml, made in transparent fused quartz instead of Pyrex, are required. Iodine flasks made of silica are particularly suitable. Otherwise, the combustion equipment is as described in Section 3.2.

Automatic pipettes are recommended for the repetitive dispensing of reagents. The Linson pipette has been found suitable.

A spectrophotometer capable of measuring absorbance at 620 nm* equipped with cells of 4 cm path-length.

Standard flasks. It is recommended that sets of 100, 250, 500 and 1000 ml standard flasks, at least 6 of each size, should be kept ready for use in the determination.

12.2.2 Reagents

Succinate buffer solution, 0.1*M* (pH 4.6). To 2.95 g of succinic acid add about 150 ml of water. Add 20.5 ml of 1*M* sodium hydroxide from a burette (i.e.

*The Pye–Unicam model SP 600 has been found to be suitable.

81% of the amount required to form the monosodium salt), swirl until all the solid has dissolved and dilute to 250 ml. Make up fresh solution in this way each day it is needed.

Alizarin fluorine blue (AFB) is 3-aminomethylalizarin-*N,N*-diacetic acid. It is also known as alizarin complexone.

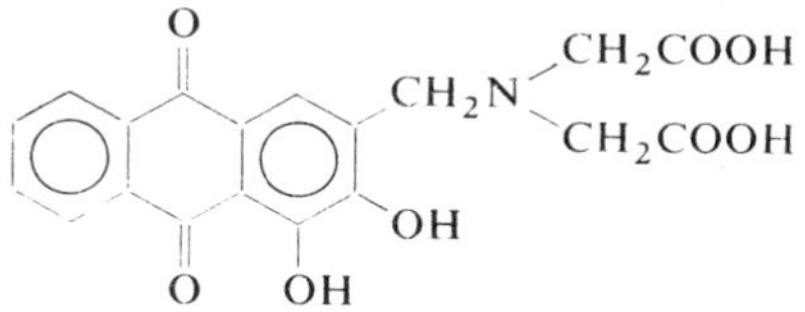

$$C_{19}H_{15}O_8N = 385.3$$

The solution used is $0.5 \times 10^{-3}M$. Weigh on a watch-glass 0.385 g of solid AFB and rinse it with a little water into a 2 litre standard flask. Take about 2 g (22 pellets) of sodium hydroxide, rinse each pellet briefly with water to remove carbonate, and drop them into 50 ml of water, to give an approximately 0.5*M* solution. Add half or a little more of this to the 2 litre flask and swirl it for about 5 min to dissolve nearly all the solid AFB. An intensely blue solution is produced. Dilute to about 1500 ml with water, add 50 ml of the succinate buffer solution and then 1*M* hydrochloric acid gradually, with mixing, until the solution is nearly yellow when viewed through the narrow neck of the flask. The AFB is its own indicator and a thin layer of the solution of the required pH (4.5–5.0) has a yellowish pink colour. Add 200 ml of propan-2-ol (the stock of recovered propan-2-ol, p. 88, is suitable). Mix well and dilute to the mark with water. Filter before use. If stored at about 0°C, it will keep for about six months but regular recalibration is essential for the most accurate results.

Lanthanum nitrate solution, $0.9 \times 10^{-3}M$. Dissolve 0.80 g of lanthanum nitrate [$La(NO_3)_3.6H_2O$ = 443.0] in water and dilute to 2 litres (mark on polythene bottle). (Note that 5 ml of this solution is to be taken and 10 ml of the AFB solution, so there is theoretically a 10% excess of AFB, to allow for the AFB not being absolutely pure.)

Propan-2-ol, 50% solution. Mix 200 ml of propan-2-ol with 200 ml of water. As heat is generated this is best done early on the day in which it is to be used. (The propanol recovered as described in Chapter 8, p. 88, may be used).

Perchloric acid, 2*M*. Dilute 20 ml of perchloric acid, s.g. 1.54, to 100 ml with pure water.

12.2.3 Procedure

A calibration graph should be prepared by making up a series of standard solutions of sodium fluoride, and applying the following procedure to 10 ml aliquots. It is suggested that the graph cover the range 8–16 μg of fluoride.

Rinse with pure water as many 100 ml standard flasks as will be required, suitably labelled, and put into each 50 ml of the 50% propanol-water mixture. This will produce a final solution containing 25% propanol, but avoids the rise of temperature that occurs when propanol is mixed with water. Add to each a 10 ml aliquot of the appropriate solution. To an additional 100 ml flask add 10 ml of pure water to provide the reference solution. Add 10 ml of succinate buffer to each, followed by 10 ml of AFB solution (freshly filtered) and 5 ml of lanthanum nitrate solution, mixing well after each addition. Automatic pipettes may be convenient. Dilute the solutions to the mark with pure water after mixing well. At this stage the reference solution is slightly bluish-pink, but the other solutions, if at least 10 μg of F is present, will be bluer. Allow the solutions to stand for 30 min in the dark.

Meantime, switch on the spectrophotometer, then when the solutions are ready, measure their absorbances in a 4 cm cell, against the reference solution in a matching cell.

When a compound is to be analysed for fluorine, the analysis should be done in duplicate, and a standard fluorine compound in which the fluorine is bound similarly (e.g. in $-CF_3$ groups) should be analysed at the same time. *Minor* adjustments may then be made to the calibration graph.

Take sufficient sample to give about 12 μg of fluoride in the final 10 ml of solution. For example, take 6 mg of a sample containing 10% of fluorine and dilute to 500 ml; 10 ml of this solution will then contain $6 \times 1/10 \times 1/50 = 0.012$ mg (12 μg) of F.

Combust the unknowns and calibration substances one after another in the usual way, except that the absorption solution is 20 ml of pure water. Allow at least 10 min for absorption.

Remove the stopper and rinse it and the platinum into the flask thoroughly with pure water. Add 1 drop of 2*M* perchloric acid (2 drops if it is to be diluted to 500 ml and 4 drops if to 1 litre). Then boil the solution briefly, heating it over a flame for exactly 60 sec in all. This is to convert sulphite into sulphate, if present, and also to remove the carbon dioxide produced in the combustion. Perchloric acid does not interfere [5]. Cool to room temperature, then transfer the solution to a standard flask of appropriate size (250, 500 or 1000 ml) with plentiful rinsing with pure water, and dilute to the mark. Mix well. Check that all flasks are at room temperature. (It is an advantage to use a large container of pure water that has been stored long enough in the laboratory to be at room temperature.)

Complete the determinations as described for the calibration. For each absorbance read the quantity of fluoride from the calibration graph. Then calculate a correction factor based on the results for the standard(s): for example, a reading of 0.292 might correspond on the graph to 12.07 μg of F but if this reading were obtained for 4.491 mg of standard fluorobenzoic acid (13.56% F) combusted and diluted to 500 ml (which would give $4.491 \times 1000 \times 13.56 \times 10/(500 \times 100)$ $= 12.18$ μg of F) the correction to be applied would be $12.18/12.07 = 1.009$.

Example of calculation. Suppose 4.732 mg of sample were combusted, the products were dissolved and made up to 500 ml, and 10 ml were put into a 100 ml flask, the colour was developed and the solution made up to 100 ml. The absorbance of the 4 cm layer was 0.289. On the graph this corresponded to 11.95 μg of F. Therefore the corrected amount of F was $11.95 \times 1.009 = 12.06$ μg. Hence the amount of F in the sample was

$$\frac{12.06 \times 50}{1000} \text{ mg} ; \quad \frac{12.06 \times 50}{1000 \times 4.732} \times 100 = 12.74 \text{ \%F}.$$

If more than one standard has been analysed, the correction factors should be averaged. It has been found that this correction decreases for the first few days in the life of the AFB solution, i.e. the colour formed with fluoride intensifies, but afterwards the blue colour developed becomes slightly less intense on storage. However, an AFB solution (if stored cold) can be used in this way for about 6 months. It is possible to obtain approximate F contents without recalibration, but recalibration is advisable for more accurate results.

REFERENCES

[1] C. A. Johnson and M. A. Leonard, *Analyst,* 1961, **86,** 101.
[2] M. A. Leonard, *Alizarin Fluorine Blue,* Hopkin & Williams Monograph No. 45, 1962.
[3] R. Belcher and J. C. Tatlow, *Analyst,* 1951, **76,** 593.
[4] G. Ingram, *Methods of Organic Elemental Microanalysis,* Chapman & Hall, London, 1962, p. 213.
[5] E. J. Newman, Personal communication.
[6] R. J. Hall, *Analyst,* 1963, **88,** 76.
[7] Analytical Methods Committee of the Society for Analytical Chemists, *Analyst,* 1971, **96,** 384.
[8] R. J. Hall, *Analyst,* 1968, **93,** 461.
[9] M. Hanocq and L. Molle, *Anal. Chim. Acta,* 1968, **40,** 13.
[10] D. A. Shearer and G. F. Morris, *Microchem. J.,* 1970, **15,** 199.
[11] F. Martin and A. Floret, *Chim. Anal. (Paris),* 1958, **40,** 120.
[12] A. Steyermark, *Quantitative Organic Microanalysis,* 2nd Ed., Academic Press, New York, 1961, p. 112.
[13] H. H. Willard and O. B. Winter, *Ind. Eng. Chem., Anal. Ed.,* 1933, **5,** 7.
[14] G. M. Maciak and E. Kozłowski, *Microchem. J.,* 1979, **24,** 421.

CHAPTER 13

Sulphur

13.1 SULPHUR BY COMBUSTION METHODS [1-7]

The sample is burnt in the usual way in filter paper in an oxygen-filled flask. The absorption solution is alkaline peroxide.

After combustion the solution in the flask is neutralized, made slightly acid with hydrochloric acid and boiled, and an exactly known volume of barium chloride is added to the hot solution. The sulphate produced from the sample is precipitated as barium sulphate. The barium remaining in excess is determined by making alkaline with ammonia solution, adding enough zinc-EDTA complex to exchange the barium ions for zinc ions and titrating the latter with EDTA, with Solochrome Black as indicator. The end-point is taken when the red colour of the zinc-Solochrome Black complex just changes to bluish-green (viewed in the light from a 60 W tungsten lamp). The sulphur content is calculated from the quantity of barium chloride required to precipitate the sulphate ions.

13.1.1 General discussion

The method described has been used successfully for many years. Phosphorus and alkali metals do not interfere. It is interesting that Campbell *et al.* [8] found that phosphorus did not interfere when the barium sulphate, precipitated from acid solution, was filtered off and redissolved in an excess of EDTA, the surplus of which was determined by titration with standard magnesium chloride solution. Selig [9] avoided interference by phosphorus by titrating the sulphate potentiometrically with a lead solution, using a lead-selective electrode.

It is necessary to do the titration under electric light and experience in recognizing the end-point is necessary, but it is not difficult, with practice, and each operator should consistently take the point at which the change seems most definite. Some describe this as the point at which suddenly no red colour at all can be seen but a green colour is present. (The change is from red to blue but in the presence of Methyl Red, which is yellow because the solution is alkaline with ammonia, it is from orange-red to green.)

When an end-point seems to be reached, a further 5 ml of concentrated ammonia solution is added and the red colour (of zinc–Solochrome Black) reappears and a few more drops of EDTA solution are required to bring about the change from red to green again. As always, best results are obtained if things are done as consistently as possible. In addition, it is desirable that the back-titration be about 5 ml. As it would not seem wise to use standard solutions more dilute than 0.01*M* this means that as much as 16 mg of sample is sometimes needed (for S contents of 10% or less). If smaller amounts have to be taken the precision of the results is not so good.

13.1.2 Apparatus

Combustion flasks, and platinum baskets, as described in Section 3.2.
Burette, 10 ml, graduated at 0.02 ml intervals, with pressure filling device (for EDTA).
Burette with free piston which will deliver exactly 10 ml of 0.01*M* barium chloride automatically and repeatedly. It is filled from a reservoir placed above the burette so that it fills by gravity.

When tube (instead of flask) combustion has to be used (for volatile liquids, for example) the following additional apparatus is required.
Combustion apparatus, Belcher & Ingram type. The heater maintains the baffle chamber at 900°C and there is an automatically moving batswing gas burner which gives a very hot flame for heating the sample, with adjustable speed of travel. Provision is made for a small gas flame to heat the constriction at the outlet of the combustion tube.
Combustion tube conforming to British Standard 1428 [10]. A Type 2 combustion tube is required.
Absorption U-tube conforming to BS1428 [3].
Platinum boats, folded from platinum foil 25 x 20 mm, thickness 0.05 mm, to give walls 3 mm high and a boat 19 mm in length.
Silica tubing. Transparent fused silica with outside diameter about 8 mm and bore about 7 mm, cut into 50 mm lengths.
Magnetic stirrer and stirrer bars, PTFE-covered, 25 mm long.

13.1.3 Reagents

Hydrogen peroxide, 100 vol.
Sodium hydroxide solution, approximately 0.01*M*. Dilute in a polythene bottle 40 ml of 0.1*M* sodium hydroxide with demineralized water to a mark at 400 ml.
Alkaline-peroxide absorption solution. Make fresh for each determination because alkaline peroxide is not stable. Put 1 ml of 100 vol hydrogen peroxide into a 10 ml measuring cylinder, add 5 ml of 0.01*M* sodium hydroxide, dilute to 10 ml, and transfer into the combustion flask.
Sodium hydroxide solution, 0.1*M*.
Methyl Red solution, 0.25% in ethanol.
Solochrome Black, 0.1% (Eriochrome Black T). Weigh out 0.2 g of the solid, add

a mixture of 15 ml of triethanolamine and 5 ml of 98% ethanol (industrial methylated spirit). Shake well and transfer to a dropping bottle. Do not make larger quantities of solution, because it is not very stable (but stable for 7 months according to Diskant [6].)

Hydrochloric acid, 2M, in a dropping bottle.

Barium chloride solution, 0.01*M*. Dissolve 24.45 g of barium chloride dihydrate ($BaCl_2.2H_2O$ = 244.32) completely in about 100 ml of water. Transfer to a 10 litre glass container and dilute to the mark with demineralized water.

Ammonium chloride solution, 7%. Weigh 28 g of ammonium chloride into a polythene bottle containing 400 ml of demineralized water and mix well.

Ammonia solution, s.g. 0.900. Dispense in 5 ml quantities from a Zippette or similar piston dispenser.

Solution of zinc–EDTA. Weigh out 10 g of sequestric acid, disodium zinc salt (Koch-Light Labs. Ltd., code 2610h). Put into a 400 ml plastic wash-bottle, dissolve in a little water, fill with demineralized water and mix well.

EDTA, 0.01*M*. Dissolve 18.61 g of disodium EDTA dihydrate (A.R.) in water and dilute to 5 litres. Alternatively, use a commercial concentrate, diluted to the necessary extent.

Pure water. Distilled water is passed through an ion-exchange purifier before use.

13.1.4 Procedure

Weigh out sufficient sample to contain 1–2 mg of sulphur (e.g. 25 mg for 4% of S, 5 mg for 30% of S). Combust in the oxygen flask as described in Section 3.2, absorbing in the alkaline peroxide solution. Leave for at least 10 min for absorption to be complete. Remove the stopper and rinse it and the platinum thoroughly with pure water so that the total volume of the solution in the flask is about 50 ml. Add 2 drops of Methyl Red solution and just neutralize (if necessary) with 0.01*M* sodium hydroxide, added dropwise. Then add 2 drops of 2*M* hydrochloric acid. Heat to boiling on a hot-plate, boil for about 1 min, then (whilst the solution is still at above 90°C) add exactly 10 ml of 0.01*M* barium chloride. Boil for 3 min on the hot-plate, then leave to cool in an ice-bath to below 10°C. Add 2 ml of 7% ammonium chloride solution, 5 ml of ammonia solution and 2 ml of zinc–EDTA solution. Add 2 or 3 drops of Solochrome Black solution and 2 or 3 drops of Methyl Red solution (the number of drops of each indicator may be adjusted to suit the operator's colour vision), then add a stirrer bar. At this point the temperature of the solution should be between 13 and 17°C. Titrate the zinc with EDTA with a 60 W lamp lighting the solution from behind. As the end-point is approached the orange-red colour changes to a grey-blue. At this point, add a further 5 ml of ammonia solution. Titrate finally to the first appearance of a bluish-green colour (a combination of the blue of Solochrome Black with the yellow of alkaline Methyl Red).

Calculation. If y ml is the volume of EDTA solution equivalent to 10.00 ml of barium chloride, and v ml is the volume of EDTA required for titration of a combusted sample, then

$$\% \text{ sulphur} = \frac{(y-v) \times 32.06}{w}$$

where w mg is the weight of sample taken. An average value for y is obtained by combusting and titrating 2 or 3 standard compounds and calculating y from the sample weight and known amount of sulphur present. It should be noted that the weight of sample taken is chosen to ensure a back-titration volume (v) of about 5 ml.

Procedure for volatile liquids. (Weighing in glass capillary followed by tube combustion.) Switch on the combustion-tube furnace well in advance so that the combustion chamber is up to temperature (900°C). The absorber will have been left attached. Turn on the oxygen flow at 50 ml/min. Weigh out the sample in a sealed Pyrex capillary in the usual way. Detach the absorber and thoroughly rinse it with pure water, and wash the joint. Put 10 ml of alkaline peroxide solution into it. Fill the cup above the absorber with pure water, holding it there with the long handled stopper. With the aid of a wire hook, place the sample in the combustion tube inside and at the centre of the silica sheath in such a position that the centre is about 35 mm from the electric heater. Replace the bung in the combustion tube and check that the oxygen flow-rate is 50 ml/min. Light the small burner at the absorber end of the combustion tube, to prevent the condensation of sulphur trioxide at this spot. Carry out the combustion first by letting a small fish-tail flame (gas half on) traverse the sample automatically during about 3 min in order to decompose the sample gradually. Then withdraw the flame a little so that its centre is about 35 mm from the electric furnace (mark on apparatus) and heat the capillary for 7 min (alarm timer) with the gas and air full on to give a strong roaring flame. Then turn out both flames, and leave the oxygen flowing for a further 3 min. Remove the bung from the combustion tube, and by blowing and sucking by mouth along the flexible tubing attached to the splash-bulb of the absorber, agitate the absorber solution so that the entire absorber is rinsed from the Y-joint of the combustion tube to just below the ground joint of the absorber. Run the contents of the absorber into a clean 100 ml conical flask. Close the bottom tap and run the water from the upper cup into the absorber. Remove the absorber bulb and add about 5 ml of water to it. Use this water to rinse the absorber (blow and suck by mouth and tubing as before). Continue thorough rinsing of the absorber until a total of about 25 ml has been collected in the 100 ml conical flask. All the sulphur which was present in the sample should now be present as sulphate ions in this flask. Continue as for solid samples, commencing at 'Add 2 drops of Methyl Red solution . . .',

Samples containing metals. The presence in samples of metals other than sodium and potassium, e.g. magnesium, may interfere when flask combustion is used. Samples containing such metals can sometimes be analysed successfully by tube combustion, if tungstic oxide is added before the combustion. All the sulphur will be expelled and will appear in the absorber but metals such as Ca, Mg, Zn will be retained by the tungstic oxide. Many such metals would be complexed by the EDTA solution and would therefore cause low sulphur figures if flask combustion were used.

13.1.5 Alternative titrimetric finish [11]

The method given above (addition of a known amount of barium chloride in excess and back-titration with EDTA) has proved highly satisfactory for many years, and is easy to use. However the following method has also been very widely used and therefore it is given here with essential details.

The substance is combusted as already described and the solution produced (not more than 25 ml) is neutralized in the same way with a few drops of sodium hydroxide solution, boiled to destroy peroxide and then made just acid with hydrochloric acid. About 5 drops of nitchromazo solution (0.1% in ethanol) are added, followed by 100 ml of ethanol or propan-2-ol. The solution is then titrated with aqueous 0.02*M* barium perchlorate solution until the first appearance of a red colour indicates the presence of barium ions in excess. The method has the disadvantage that phosphorus interferes, although less with nitchromazo than with thorin [12] (the indicator used previously), because nitchromazo can be used in a more strongly acidic solution. Several other useful indicators are available [13,14], and an individual may well find that he will have a preference for a particular colour change.

13.2 SULPHUR BY THE MICRO-CARIUS METHOD [15-18]

The sample (not less than 20 mg) is decomposed by heating at about 300°C with 0.3 ml of fuming nitric acid in a sealed glass tube which also contains 50 mg of sodium chloride. The contents are dissolved in water, a little hydrochloric acid is added and the solution is evaporated to dryness on a steam-bath to remove nitric acid. The residue is dissolved and the solution filtered. Barium sulphate is precipitated from the boiling solution and collected, ignited and weighed on asbestos in a platinum Gooch crucible. As much as 40 mg of sample may be taken when the sulphur content is low.

13.2.1 Apparatus

Much of the apparatus required is listed in Section 3.1.

Three platinum Gooch crucibles with platinum lids. These are small platinum crucibles, the base of which is perforated with about 40 small holes. Height 30 mm, 19 mm in diameter at top and 13 mm in diameter at base. The lids and crucibles are identified by punched letters. Total weight should be 9-10 g.

Aluminium cooling block and holder for the platinum crucibles. This consists of a 3 in. diameter cylindrical block of aluminium, with 3 holes 20 mm deep and 18 mm in diameter. A rim holds a 75 mm Petri dish as a cover.

Gooch crucible adaptor for use with a 500 ml Buchner flask.
Bulb pipette, 1 ml, for barium chloride solution.
Safety pipette for nitric acid (1 ml graduated at 0.01 ml intervals).
Steam-bath with a 3 aperture plate to fit 25 ml squat-form beakers.
Platinum-tipped crucible tongs or forceps for handling the crucibles.
Filter funnel, 25 mm diameter with small adjustable stand.
Filter paper, Whatman No. 1, 50 mm.
An electric bunsen burner.

13.2.2 Reagents

Nitric acid (fuming, s.g. 1.5), A.R.
Sodium chloride. The analytical-grade reagent is made up into tablets of approximately 50 mg each (diameter 3 mm, length 3.5 mm).
Barium chloride solution, 10%. Barium chloride dihydrate, 100 g, is dissolved in a litre of water, 1 ml of conc. hydrochloric acid is added, and the solution filtered, after standing, into a smaller flask fitted with bulb pipette for easy delivery of drops.
Gooch crucible asbestos, M.A.R.
Concentrated hydrochloric acid, A.R. kept in a dropping bottle for convenience.
Ethanol. Freshly filtered industrial methylated spirit is always added from the filter funnel.
Water. Distilled and demineralized.

13.2.3 Procedure

Rinse sufficient Carius tubes with distilled water and dry in the oven. Add 0.3 ml of fuming nitric acid to each Carius tube, when they have cooled, from a safety pipette. Weigh the sample in a weighing tube (p. 28) that already contains a sodium chloride tablet. Take about 20 mg of sample if the sulphur content is ~15% and more in proportion if it is lower, but not more than 40 mg. The weight of barium sulphate obtained need rarely be less than 20 mg if enough sample is available. Drop the tube and sample into the Carius tube and seal as described in Section 3.1. Cover the seal with soot from a luminous flame, and label the tube with a letter drawn in the soot. Heat the tubes at 280–300°C for 3 hr (but shorter heating periods, even less than 1 hr, have often proved successful in cases of urgency). After cooling, open the tubes (p. 30) and dissolve the contents in a minimum of distilled water. Pour into a 25 ml squat-form beaker and follow with four small rinses of the tube. Rinse the weighing tube and remove it. The total volume of the solution need not exceed 15 ml if care is taken. Place on the steam-bath to evaporate to dryness. When dry (drops condensed on walls of beaker must evaporate), cool, dissolve the residue in the minimum of water (about 4 drops), warming it for a few sec, add 3 drops of hydrochloric acid and evaporate to dryness. Repeat this treatment to remove all traces of nitric acid. Dissolve the residue in about 10 ml of water, add 1 drop of conc.

hydrochloric acid and filter through a 25 mm conical filter into a 25 ml tall-form borosilicate glass beaker. Filtration is more rapid if the solution is hot. It is usually convenient to heat for a few min on the steam-bath. Wash the filter with several small washes until the total volume is about 18 ml. Stir and heat to as high a temperature as possible on the steam-bath and slowly add 4 drops of barium chloride solution; when adding the barium chloride do not scratch the inside of the beaker. Wait for 1 min, then slowly add 6 more drops of barium chloride solution. Leave on the steam-bath for 1 hr, then add 1 ml of barium chloride solution. This treatment is intended to avoid co-precipitation. The first 10 drops provide approximately the theoretically required quantity of barium chloride. Filter and collect the barium sulphate on asbestos in a platinum Gooch crucible, prepared as described below. The Gooch crucibles, once prepared, are used repeatedly without emptying until the filtration rate becomes too slow. Thus the weight found on completion of one determination may be carried forward to the next (though it is best to check).

To prepare the crucibles, pour a suspension of asbestos into the crucible, apply gentle suction, press down a little, add some more asbestos, and wash thoroughly. Dry and ignite as described below. The filter is most reliable when some barium sulphate has been collected on it. Two of the crucibles are used for collecting barium sulphate and the third as the counterpoise. Transfer the barium sulphate to the crucible with the aid of the wash-bottle jet and alcohol (dispensed through a filter). Complete transfer is often not easy, but the alcohol helps. Wash well with alcohol and water alternately until the beaker is clean. Finally wash the Gooch crucible with alcohol. Place the crucibles complete with lids in an oven at 100–150°C for 5 min, then transfer them to an electric 'bunsen burner' set fully on (red heat) for 10 min. Once the counterpoise has been bedded with asbestos, it is not wetted again, but it is ignited every time. Put the three crucibles into the cooling-block and weigh exactly 10 min later (while waiting adjust zero of balance). When not in use the crucibles are kept in the cooling block, with glass cover, so that they are ready for use and the weight previously found can be used again. Weigh the Gooch crucibles without their lids (these are used to protect the crucible during heating). $BaSO_4$ contains 13.73% of S. Calculate the sulphur content of the sample from this.

13.3 OTHER METHODS

Slanina *et al.* [19] have described an automated reductive method for traces of sulphur, based on hydrogenation and titration of the hydrogen sulphide evolved. Maciak *et al.* [20] have based an automated method on oxidative tube-combustion and photometric titration.

REFERENCES

[1] R. Belcher and G. Ingram, *Anal. Chim. Acta,* 1952, 7, 319.
[2] G. Ingram, *Mikrochim. Acta,* 1956, 893.

[3] S. J. Clark, *Quantitative Methods of Organic Microanalysis,* Butterworths, London, 1956, pp. 101-108.
[4] B. B. Bauminger, *Trans. I.R.I.* 1956, **32**, 218.
[5] BS903: Part B6: 1958 (The determination of sulphur in rubber).
[6] E. M. Diskant, *Anal. Chem.*, 1952, **24**, 1856.
[7] J. F. Alicino, *Microchem. J.* 1958, **2**, 83.
[8] A. D. Campbell, D. P. Hubbard and N. H. Tioh, *Mikrochim. Acta,* 1975 **II**, 209.
[9] W. Selig, *Mikrochim. Acta,* 1975 **II**, 665.
[10] BS1428: Part A5: 1965.
[11] H. Wagner, *Mikrochim. Acta,* 1957, 19.
[12] J. S. Fritz and S. S. Yamamura, *Anal. Chem.*, 1955, **27**, 1461.
[13] B. Budesinsky, *Anal. Chem.*, 1965, **37**, 1159.
[14] E. E. Archer, D. C. White and R. Mackison, *Analyst,* 1971, **96**, 879.
[15] J. B. Niederl and V. Niederl, *Micromethods of Quantitative Organic Elementary Analysis,* Wiley, New York, 1938, Chapter VII.
[16] G. Ingram, *Methods of Organic Elemental Microanalysis,* Chapman & Hall, London, 1962, Chapter 5, p. 224.
[17] A. Steyermark, *Quantitative Organic Microanalysis,* Academic Press, New York, 1961, Chapter 10.
[18] E. P. Clark, *Semimicro Quantitative Organic Analysis,* Academic Press, New York, 1943, Chapter VI.
[19] J. Slanina, P. Vermeer, J. Agterdenbos and B. Griepink, *Mikrochim. Acta,* 1973, 607.
[20] G. M. Maciak, P. W. Landis and E. Kozłowski, *Microchem. J.*, 1979, **24**, 64.

CHAPTER 14

Metals

The metals which probably occur most frequently in organic compounds are sodium and potassium. These are dealt with in the first part of this chapter, together with other metals which can be determined in the same way, that is by the sulphated ash method.

In the second part of the chapter well-tried methods for some other commonly occurring metals are described. In those cases where the author has no direct experience, a suitable method may be available in the literature [1-3].

Reference is also made to the determination of metals when they are present as traces only.

14.1 THE DETERMINATION OF SODIUM OR POTASSIUM

The sample is weighed in a platinum boat provided with a platinum sheath and the boat and sample are heated in the platinum sheath in a stream of oxygen, which contains sulphur dioxide, until only sodium (or potassium) sulphate remains in the platinum boat and sheath, where it is weighed. The sulphur dioxide reacts to produce sulphur trioxide and hence the alkali metal sulphate by the catalytic action of the hot platinum in an excess of oxygen, as described by Martin [4].

The sulphated ash must finally be heated strongly enough to decompose any bisulphate and form the neutral sulphate (e.g. Na_2SO_4).

It must be noted that in the use of this method it is necessary to know with certainty which metal is present and also to assume that it is the only metal present. If, for example, a sample stated to be a sodium salt was in fact a potassium salt then the result obtained would be erroneous (although it would be of some value because it would not confirm the theoretical value).

Calcium, barium, etc. can be determined in the same way although other methods are available (e.g. the EDTA method, p. 115).

14.1.1 Discussion of the method

It is surprisingly difficult to combust a metal-containing organic substance in an open boat or small crucible without loss of the metal to be determined. Usually the sample will melt, darken and begin to froth, so that heating must be discontinued for fear of the sample frothing over the edges. Thus, it has been found advantageous to heat the boat inside a sheath [5,6]. Also, the molten sulphate (Na_2SO_4, m.p. 800°C) has a tendency to creep and spread over the entire surface of the boat, inside and out, so that losses might result if the sheath were not used.

As already mentioned, this method is entirely unspecific and depends completely on the certainty that the only metal present is the one stated. More specific methods, e.g. determination of sodium as sodium zinc uranyl acetate [7], can be applied if necessary after a suitable destruction of the organic part (e.g. by Carius digestion). On the other hand, in the synthesis of many organic substances it is often possible to say that sodium, for example, was the only metal used. Sometimes, which is even more definite, a sodium salt of an acid has been made and needs to be analysed.

Barium can be determined by the sulphated-ash method (the residue, $BaSO_4$, contains 58.9% of Ba) but it is more specific to precipitate it from solution with an excess of sulphuric acid after, e.g., a Carius destruction (unless sulphur is present in the sample, in which case barium sulphate would be formed in the Carius tube).

If a metal is present in small amounts as an unintentional impurity the sulphated ash may be determined in the same way in the platinum boat and sheath. For example 0.2 mg of ash from 20 mg of sample would be weighable with a useful precision and would represent 1% of ash or approximately 0.3% of sodium.

14.1.2 Apparatus

Platinum boat and sheath. As described in BS1428 [8]. A suitable boat may also be made from platinum foil as described on p. 43. The sheath can also be made from platinum foil of the same thickness (0.05 mm) by cutting a rectangle 35 mm × 27 mm, rolling it round a metal rod of 6 mm diameter ($\frac{1}{4}$ in.) and forming a seam of 4 thicknesses along its length and hammering this while red hot. A wooden former is used not only to shape the sheath when making it but also to re-form it if it becomes distorted in use. It produces a flat-bottomed tunnel shape. The wire handle shown in the standard is not really necessary but can easily be welded on by hammering at red-heat. For quick weighing a counterpoise should be made and its weight adjusted to match that of the boat and sheath combined. It may be made from scrap platinum or non-magnetic wire.

Glass bubbler, to contain the sulphur dioxide solution.

Combustion tube, fused quartz, 13 mm outside diameter, and about 250 mm long.

The apparatus is set up in a fume cupboard because of the sulphuric acid fumes produced.

Bunsen burners, (a) micro burner [9] and (b) burner with flame spreader.

14.1.3 Reagents

Sulphur dioxide gas in a 500 g canister.

14.1.4 Procedure

Before preparing the platinum boat and sheath and weighing the sample, prepare fresh saturated sulphur dioxide solution by bubbling the gas into water in the glass bubbler, which is already connected to the combustion tube with a rubber bung. The platinum boat and sheath are stored in dilute nitric acid when not in use. Take the boat and sheath out of the acid, rinse with pure water and, holding them with platinum-tipped tweezers, heat to red heat in a flame, then stand them on a metal block for 10 min before weighing against the counterpoise. Add the sample to the boat and reweigh. For sodium or potassium take about 30 mg if the metal content is expected to be below 10%, but 5-10 mg if it is expected to be above 20%. Too much sample should be avoided or losses may occur owing to frothing and creeping. Aim at about 5 mg of sulphated ash. Disconnect the sulphur dioxide canister from the bubbler and connect the oxygen supply to the bubbler and pass the gas at about 2 bubbles a second so that sulphur dioxide-laden oxygen is passing into the combustion tube. Put the platinum boat, placed centrally in the sheath, into the other end of the combustion tube, so that the end of the sheath is exactly at the end of the quartz tube.

Begin to heat the platinum through the quartz tube by means of a microburner flame. The first stage may be the distilling away of much of the organic part of the substance. Care should be taken during the first stages; a very small flame directly under the quartz tube may be all that is needed. The flame should slowly be increased but it need not occupy all the attention of the operator, who could also be doing other things. After about 30 min, the micro burner should have been full on for about 10 min. Change to the ordinary burner with flame spreader and keep the tube just red hot for a further 10 min. A short electric tube heater can be devised, the temperature of which will rise from room temperature to red heat in about 30 min, thus effecting a gradual ashing of the sample; however the simple gas flames are quite effective.

Withdraw the platinum sheath and boat with platinum-tipped tweezers and heat in the spread flame to red heat for about 1 min. The hottest flame possible with such a burner should be avoided because sodium sulphate may begin to volatilize. Then cool the boat and sheath on a metal block for 10 min and weigh as before. The increase in weight is due to the metal sulphate. Sodium sulphate contains 32.4% of Na and potassium sulphate contains 44.9% of K. Calculate the metal content of the sample accordingly. Repeat the heating and cooling and reweigh to confirm constancy of weight.

14.2 THE DETERMINATION OF COPPER [10]

Sufficient sample (but not more than 50 mg) to contain 7–10 mg of copper is digested in a sealed micro-Carius tube by heating for about 3 hr at 300°C, with 0.3 ml of fuming nitric acid.

After dilution the copper solution is made alkaline with a little sodium hydroxide solution and then just acid with dilute acetic acid; then about a 3-fold excess of a cold aqueous solution of salicylaldoxime is added to the cold solution slowly with stirring.

The pale green copper complex is collected, washed with cold water, dried at about 100°C and weighed.

14.2.1 Discussion of the method

In a laboratory where the determination of copper is required very rarely indeed (once or twice in a year) and the micro-Carius equipment is available, this simple method has proved easy and reliable. Smaller quantities than those recommended may be used but if plenty of sample is available the amounts recommended are best, to minimize errors in a method not used often enough to be highly perfected.

Although the precipitate is formed at room temperature it can always be filtered off and washed rapidly.

14.2.2 Apparatus

Carius apparatus, as described in Section 3.1.1.
Pyrex weighing tubes, type (b) in Section 3.1.1.
Pyrex beakers, 50 ml.
Glass stirring rods, 80 mm long, 2.5 mm in diameter.
Sintered-glass crucibles, 4 cm tall. Three crucibles are used together, one as a counterpoise. They are cleaned for use by immersion in aqua regia (warm if necessary). When washed with water and dried they should return to their original weight, or nearly so.

14.2.3 Reagents

Fuming nitric acid A.R.
Sodium hydroxide solution, 2M.
Dilute acetic acid solution, 2M.
Salicylaldoxime, aqueous solution, 1%. If salicylaldoxime is not available the solution can readily be made from salicylaldehyde as follows. Dissolve 1.22 g of salicylaldehyde ($C_7H_6O_2 = 122$) in 5 ml of industrial methylated spirit (solution *a*). Dissolve 0.7 g of hydroxylamine hydrochloride in 1 ml of water and add 8 ml of industrial methylated spirit (solution *b*). Add *a* to *b*, warm to 40°C and let stand for about 30 min, then add 120 ml of hot water (approximately to a mark on the label of the bottle) at 80°C. Let cool before use. This solution keeps for only three days. One ml reacts with about 2.6 mg of copper.

14.2.4 Procedure

Wash the weighing tubes with pure water, dry them in the oven and leave in the balance room under cover at least overnight before use. Add a wisp of cotton wool to make a barrier to support the sample at the constriction. Wash the Carius tubes with pure water, allow to drain, dry in the oven, cool and add 0.3 ml of fuming nitric acid from the safety pipette. Weigh out sufficient sample to contain 7–10 mg of copper, but not more than 50 mg of sample. Put the sample in its weighing tube into the Carius tube and seal, keeping the sample out of the acid until the tube is sealed, if possible. Heat the sealed tube for $3\frac{1}{2}$ hr. After cooling, open the tube as described on p. 103, rinse with pure water into a 50 ml beaker, with dilution to about 25 ml, transferring the weighing tube to the beaker. If iodine is present heat on a steam-bath until all the iodine has evaporated, i.e. the solution is colourless or only pale blue (pure copper nitrate). Remove the weighing tube with rinsing, add dilute sodium hydroxide solution dropwise until copper hydroxide just separates and then add 5 ml of 2*M* acetic acid. Add very slowly, with stirring, 10 ml of salicylaldoxime reagent, freshly filtered. After letting stand for about 30 min, transfer the pale green precipitate to the weighed sintered-glass crucible, using cold water to rinse the beaker and wash the precipitate. Dry for 30 min at a temperature not exceeding 105°C. Weigh after cooling for 30 min in the desiccator. The precipitate contains 18.94% copper. From the weight of precipitate calculate the percentage of copper in the sample.

14.3 THE DETERMINATION OF GOLD

If no metal other than gold is present, the gold will remain in the platinum boat when a carbon and hydrogen determination is carried out, and may be weighed.

When another metal, such as sodium, is present, the weighed sample in the platinum boat is covered with tungstic oxide, and a carbon and hydrogen determination done by combustion in a stream of oxygen. The boat with residue is then immersed in 30% sodium hydroxide solution in a 10 ml beaker and digested on a steam-bath. The solution is decanted and the residue and boat are well washed with water. This procedure dissolves away the tungstic oxide and any soluble salts. The boat and residue are dried and weighed to give the weight of the gold. The gold tends to form a skeleton which adheres to the platinum by partial alloying so the washing described causes no loss of gold.

This method, which has been found useful for the drug gold sodium thiomalate, could be applied to similar compounds.

14.4 THE DETERMINATION OF IRON

As much as possible (up to 50 mg) of the iron compound is digested in a Carius tube with nitric acid. After dilution, the iron is precipitated with ammonia, and the precipitate is washed, dissolved in nitric acid, and finally ignited and weighed as ferric oxide (Fe_2O_3) in a small (1 ml) platinum crucible.

14.4.1 Discussion of the method

If iron compounds are submitted for analysis only very rarely, this reliable straightforward method is very convenient if micro-Carius equipment is available. Because the ferric oxide which is the final weighing form contains 69.9% of iron, it is advisable to take as much sample as can be spared. However, reasonable precision is not usually difficult to achieve. Any organic compound which contains no metal other than iron can, of course, be ignited (in a platinum crucible for preference) with additions of nitric acid, and weighed as ferric oxide, but great care is needed to avoid losses by frothing.

The micro-Carius procedure is preferable because it avoids losses due to frothing and also will separate iron from some other metals (e.g. sodium) if they are present.

14.4.2 Apparatus

Centrifuge tube, graduated up to 10 ml (total volume to the mouth, 13 ml).
Teat pipette. A standard disposable Pasteur pipette with a 1 ml plastic teat.
Platinum crucible, 1 ml, and lid.
Small silica triangle formed from heavy-gauge Kanthal wire threaded into 3 pieces of silica tubing, each 17 mm long so that a 1 ml platinum crucible can be supported in it at about half height.
Centrifuge, suitable for 10 ml tubes. A manual model is convenient.

14.4.3 Reagents

Ammonia solution, 50%. Dilute concentrated ammonia solution with an equal volume of water.

14.4.4 Procedure

Take about 50 mg of sample if possible (compress it thoroughly in the weighing tube because some iron compounds, e.g. ferrocenes, may explode in the nitric acid vapour before the tube is sealed) in the Carius tube, and proceed as for the determination of phosphorus (p. 136) up to the heating and cooling of the tube. Open the cold Carius tube as described on p. 103. When the tube is cold rinse the contents into a 10 ml centrifuge tube to a total volume of about 7 ml. Add a little concentrated hydrochloric acid if this is necessary to obtain complete dissolution, but if necessary filter off any insoluble matter still visible. Warm the solution in the centrifuge tube and add, with stirring with a glass rod, an excess (but not too great) of 50% ammonia solution. Check that an excess of ammonia is present, then heat the tube in the steam-bath for

10 min (gently at first as some ammonia will boil off). Centrifuge the tube while hot, and wash the precipitate twice with 3 ml of water from a teat pipette. Dissolve the precipitate in the minimum of nitric acid and transfer the solution by means of the teat pipette to the 1 ml platinum crucible (previously ignited and weighed). Evaporate to dryness on the steam-bath, supporting the crucible on the small silica triangle, then dry in an oven at 150°C for 10 min. Finally ignite the crucible and precipitate at 750°C for at least 5 min. (For this stand the crucible on its lid inside a porcelain crucible.) Cool on a brass block and weigh 5 min after removal from furnace. The residue is assumed to be Fe_2O_3, which contains 69.9% of Fe.

14.5 THE DETERMINATION OF MERCURY

Mercury may be determined gravimetrically as the sulphide, after Carius digestion in nitric acid.

14.5.1 Discussion

If mercury has to be determined only very rarely (say once a year) this method is suitable if the laboratory already has Carius apparatus and experience. There is a slight risk that the mercuric sulphide may contain some elementary sulphur but even an approximate result will allow the characterization of the mercury compound (e.g. whether the molecule contains one or two mercury atoms). The Carius procedure has a definite advantage over open-vessel methods in that no losses of the rather volatile mercury and its compounds can occur. When a better method is required for some more frequent use the method of Wickbold [11], as described by Roth [12], is recommended.

The sample (2-5 mg) is destroyed in a sealed Carius tube at 250°C, but 0.5 ml of ordinary conc. nitric acid (65%) may be used. The contents of the tube are diluted and the mercury is titrated with 0.01*M* solution of sodium diethyldithiocarbamate [2.253 g of the trihydrate $(C_2H_5)_2NCS_2Na.3H_2O$, per litre]. This solution is not stable so it must be standardized each time against 0.01*M* mercuric sulphate under exactly the conditions used for the sample. The Carius tube is rinsed into a 250 ml conical flask (fitted with a stopper) with about 35 ml of water. Tartaric acid (10 ml of 10% solution), 1 ml of cupric acetate solution (0.1% in 1% w/v sulphuric acid solution), sufficient 25% ammonia solution to render the solution definitely ammoniacal (pH about 9), and 5 ml of chloroform are added. The solution is then titrated with the 0.01*M* dithiocarbamate, the white precipitate of mercuric dithiocarbamate first formed being partly dissolved, on shaking, by the chloroform layer. The end-point is indicated by a permanent yellowish-brown colour in the chloroform, due to copper diethyldithiocarbamate. Each ml of the 0.01*M* solution corresponds to 1.003 mg of mercury.

The more traditional micro-method, based on that of Boëtius [13], in which the sample is decomposed in a tube-combustion and the liberated mercury is collected and weighed as an amalgam on gold wool or foil, is too elaborate for very occasional use but might be justifiable for frequent use.

In an interesting method that is widely used for some pharmaceutical substances (e.g. mercurochrome), the mercury is reduced by zinc and forms an amalgam with it, and can then be titrated with standard ammonium thiocyanate solution after dissolution in dilute nitric acid.

14.5.2 Apparatus and reagents as for other Carius methods (p. 27). A source of hydrogen sulphide (preferably a cylinder).

14.5.3 Procedure (sulphide method)

Digest 40–50 mg of the sample, if sufficient is available, in a sealed Carius tube in the usual way with 0.3 ml of fuming nitric acid at about 280°C. Heating for 2 hr is usually sufficient. Since the atomic weight of mercury is high (200.6), large amounts (up to 50 mg) of the sample can be heated in the usual Carius tube because the proportion of organic material to be destroyed will be low.

Rinse the contents of the tube, after cooling and opening, into a small glass evaporating dish and evaporate just to dryness on a steam-bath to remove most of the residual nitric acid. Do not prolong the heating when dry or mercury may be lost by volatilization.

Dilute to about 50 ml in a 150 ml beaker with the addition of about 5 ml of 2*M* hydrochloric acid, and bubble hydrogen sulphide into the solution through a narrow glass tube. If mercury is present a brown precipitate of mercuric sulphide will separate. Saturate the solution with hydrogen sulphide and leave to stand for about 1 hr.

Filter off on a weighed, porosity-4 sintered-glass crucible, wash the precipitate on the filter with water and alcohol, then with carbon disulphide to dissolve any elementary sulphur. Wash with alcohol again, dry at 120°C, cool and weigh. Use a similar crucible as a counterpoise throughout (as described for other such gravimetric finishes).

The mercuric sulphide contains 86.2% of mercury, from which the mercury content of the sample can be calculated. The conversion factor is not very favourable and therefore a large sample is recommended, to provide as high a precision as possible.

14.6 THE DETERMINATION OF NICKEL

Sufficient sample to contain 5–10 mg of nickel is treated as for the determination of copper (p. 124) up to where the Carius tube is rinsed out. The solution is rinsed into a 100 ml beaker, diluted to about 50 ml, and heated to about 80°C on the steam-bath. Dimethylglyoxime (10 ml of 1% solution in ethanol) is added and then enough dilute ammonia solution to make the mixture alkaline. The red precipitate is collected in a weighed sintered-glass crucible, washed with hot water and dried at 120°C. The precipitate contains 20.32% of nickel.

14.7 SELENIUM AND TELLURIUM

The procedures described are suitable for most selenium and tellurium compounds, but the presence of ionic chloride will cause losses of Se or Te as the volatile tetrachlorides and certain metals which form very insoluble selenites or tellurites (e.g. mercury) cause low results to be obtained.

If many selenium compounds are to be determined, oxygen-flask combustion (with a quartz spiral instead of a platinum basket) is faster and may therefore be more convenient: in this case chlorine and bromine may be determined simultaneously by titration with silver nitrate at pH 4 before the selenite is titrated. The original paper [14] should be consulted for full details. Tellurium compounds cannot be combusted satisfactorily in the oxygen flask.

Other possible methods have been reviewed [15]. A method suitable for compounds containing selenium and phosphorus is given by Binkowski and Rudnicki [16]. After wet decomposition of the sample, selenium is precipitated with thioacetamide and phosphate is titrated with lanthanum.

14.7.1 Apparatus

Micro-Kjeldahl flasks and digestion rack.
Beaker, 50 ml.
Microburette. An Agla micrometer syringe burette with a sealed-in platinum wire (to act as a titrant-stream reference electrode) is very convenient.
Indicator electrode. Pure silver wire, ~1–2 mm in diameter.
Potentiometer, e.g. pH-meter.
Magnetic stirrer and stirrer bar.

14.7.2 Reagents

Silver nitrate, 0.1*M*.
Acetone.
Borate buffer. A 0.1*M* solution of borax (i.e. 0.4*M* orthoborate).
Sodium hydroxide, ~0.1*M* and ~10*M*.
Nitric acid. Concentrated and ~0.1*M*.
Sulphuric acid. Concentrated.
Universal indicator, B.D.H.

14.7.3 Selenium

Weigh a sample (4–8 mg) into a micro-Kjeldahl flask. Add 0.5 ml of conc. sulphuric acid, then heat for 60 sec with a flame just big enough to cause charring after 50 sec. Cool for 5 min, add about 5 drops of conc. nitric acid, and heat with a small flame for 15 sec only. After a further 5 min, add 3–5 ml of water, and allow to cool. Transfer the digest to a 50 ml beaker, washing the flask thoroughly with the minimum of water. Add 0.5 ml of universal indicator and 1 ml of borate buffer. Insert a magnetic stirrer bar and place on the stirrer, neutralize the sulphuric acid with 10*M* sodium hydroxide, then adjust to pH 8–8.5 by adding 0.1*M* nitric acid or 0.1*M* sodium hydroxide (solution a definite green colour). Add acetone to make the final proportion of acetone to water 1:2 v/v. Insert the silver wire and the burette tip, and titrate with 0.1*M* silver nitrate to a potentiometric end-point. If the pH drops during the titration, adjust again to 8–8.5 with 0.1*M* sodium hydroxide.

$$0.1 \text{ ml of } 0.1M \text{ silver nitrate} \equiv 0.3948 \text{ mg of Se}$$

If the compound contains a metal which forms a sparingly soluble selenite, add 3 ml of 0.01*M* iminodiacetic acid solution to the solution in the beaker before neutralizing the sulphuric acid. If a halogen is present, do the initial digestion with a mixture of 0.5 ml of conc. sulphuric acid and 5 drops of conc. nitric acid. After cooling, add two more drops of nitric acid, then proceed as before. If the compound contains Se–Se or P–Se bonds, add the nitric acid before the sulphuric acid at the start of the digestion, then proceed as before.

14.7.4 Tellurium

Weigh a sample (at least 4 mg, but more if possible) into a micro-Kjeldahl flask, then add 0.5 ml of conc. sulphuric acid and 5 drops of conc. nitric acid. Heat over a moderate flame until fumes of sulphur trioxide just appear. Cool, add a further 5 drops of nitric acid, and heat to fumes once more. Cool, then continue as for selenium from 'Transfer the digest. . .'.

$$0.1 \text{ ml of } 0.1M\ AgNO_3 \equiv 0.638 \text{ mg of Te}$$

14.8 OTHER METHODS

Bishara *et al.* [17,18] have described polarographic methods for determination of aluminium, cadmium, cobalt, copper, iron, lead, magnesium, manganese nickel, uranium and zinc in organometallic compounds, and Sakla *et al.* [19] have described the use of morpholinium morpholine-*N*-dithiocarboxylate for determination of bismuth, cadmium, cobalt, copper, iron, lead, manganese, mercury, nickel, silver, uranium and zinc.

REFERENCES

[1] R. Belcher, D. Gibbons and A. Sykes, *Mikrochemie,* 1952, **40**, 76.
[2] A. Sykes, *Mikrochim. Acta,* 1956, 1155.
[3] A. M. G. Macdonald and P. Sirichanya, *Microchem. J.,* 1969, **14**, 199; *see also* R. Belcher, B. Crossland and T.R.F.W. Fennell, *Talanta,* 1969, **16**, 1335.

[4] F. Martin, *Mikrochemie,* 1951, **36/37**, 660.
[5] H. I. Coomba, *Biochem. J.*, 1927, **21**, 404.
[6] H. K. Alber, *Mikrochemie,* 1935, **18**, 92.
[7] H. H. Barber and I. M. Kolthoff, *J. Am. Chem. Soc.*, 1928, **50**, 1625.
[8] BS1428: I1; 1953.
[9] BS1428: E4; 1962.
[10] A. I. Vogel, *Quantitative Inorganic Analysis,* 3rd Ed., Longmans, London, 1961, p. 478.
[11] R. Wickbold, *Z. Anal. Chem.* 1956, **152**, 261.
[12] H. Roth, *Pregl-Roth Quantitative Organische Mikroanalyse,* Springer, Vienna, 1958, p. 184.
[13] M. Boëtius, *J. Prakt. Chem.*, 1938, **151**, 279.
[14] M. R. Masson, *Mikrochim. Acta,* 1976 **I**, 399.
[15] M. R. Masson, *Mikrochim. Acta,* 1976 **I**, 419.
[16] J. Binkowski and A. Rudnicki, *Mikrochim, Acta.* 1977 **I**, 371.
[17] S. W. Bishara, *Mikrochim. Acta,* 1973, 25.
[18] S. W. Bishara, A. B. Sakla, M. E. Attia and H. N. A. Hassan, *Mikrochim. Acta,* 1974, 257.
[19] A. B. Sakla, A. A. Helmy, W. Beyer and F. E. Harhash, *Talanta,* 1979, **26**, 519.

CHAPTER 15

Phosphorus, arsenic and germanium

15.1 PHOSPHORUS

The substance is weighed in a special small glass weighing-tube and heated in a sealed micro-Carius tube with 0.3 ml of fuming nitric acid for about 3 hr at 300°C. About 50 mg of sodium chloride is included to prevent reaction of the phosphorus with the glass.

After cooling, the contents are diluted, hydrochloric acid is added, the solution is boiled, citric acid is added to prevent interference by silica, and the phosphoric acid is precipitated as quinoline phosphomolybdate by the addition of quinoline molybdate solution [1].

The yellow precipitate is collected on a filter paper in a small filter and washed with water. An excess of standard sodium hydroxide (20 ml of 0.1*M*) is added and after complete dissolution of the yellow precipitate, the excess of alkali is titrated with 0.1*M* hydrochloric acid to a phenolphthalein end-point.

15.1.1 Discussion

The method described has for many years proved very reliable and easy to carry out when Carius equipment is available in the laboratory. Because 1 atom of phosphorus gives rise to 26 equivalents of acid (2 from the first two neutralization steps of phosphoric acid and 24 from the 12 molybdic acid molecules derived from the heteropoly acid) 2 mg of phosphorus is equivalent to about 17 ml of 0.1*M* sodium hydroxide. Such quantities can be measured very accurately and have given excellent results, but they may be greatly reduced if the need arises. (Belcher [2] has used as little as 70 μg of sample, containing e.g. only 7 μg of phosphorus, which would require about 30 μl of 0.02*N* sodium hydroxide for its titration.)

The Carius method of destruction has been criticized because it is thought that some phosphorus is lost by reaction with the glass of the Carius tube [3]. Such losses, if they do occur, are reduced by the presence of an alkali metal ion [4] and this is done here by use of the sodium chloride pellets recommended for the Carius sulphur determination.

Oxygen-flask combustion is perfectly satisfactory in most cases provided that an oxidizing agent is used in the absorption solution. Alkaline hypobromite is recommended. However, flask combustions of quinine hypophosphite were found by Belcher and Macdonald [5] to give low results. On the other hand the Carius oxidation has been found by the author to be perfectly satisfactory with quinine hypophosphite and therefore, presumably, with all hypophosphites. In the oxygen-flask combustion of arsenic compounds, a platinum sample holder is not suitable because some arsenic may be lost on the platinum. The Carius oxidation of arsenic compounds is excellent, although care must be taken not to use too much citric acid, (added to avoid high results due to silica), and an additional treatment with potassium chlorate is necessary.

The author has found the simple filter with siphon tube both neat and rapid although the Witt plate and paper-pulp pad used by Belcher and Macdonald [5] is probably a little more rapid.

The formula for quinoline phosphomolybdate is $(C_9H_7N)_3H_3PMo_{12}O_{40}.2H_2O$ from which can be seen that the quinoline molybdate reagent (QM) should contain a 4:1 molar ratio of molybdenum trioxide (m.w. 144) to quinoline (m.w. 129). In the QM reagent described, 150 g of molybdenum trioxide and 28 ml of quinoline (s.g. 1.095) are used, roughly the correct proportions [5].

When the reagent is used for determination of phosphate, 31 mg of phosphorus would require $12 \times 144 = 1728$ mg of molybdenum trioxide. Therefore about 2 mg would require about 112 mg. The QM reagent contains about 10% of molybdenum trioxide so the 10 ml recommended contains about 1 g (i.e. approximately a 9-fold excess). Thus although the reagent has often been found to deposit crystals on storage there is still adequate reagent in solution when it is filtered, and the recommended 10 ml is used. Alternatively it may be warmed to redissolve the crystals.

The yellow precipitate has been weighed as the dihydrate after drying at 160°C [6].

In the past, ammonium phosphomolybdate was similarly used, but it was never regarded as stoichiometric and an empirical factor had to be used [7]. However, Floret [8], following the hypothesis of Posternak [9], has shown that the theoretical weight of ammonium phosphomolybdate is obtained if all mineral acids (e.g. sulphuric acid) which contain oxy-ions are excluded when the ammonium compound is precipitated. Thus, hydrochloric acid may be present, but sulphuric and nitric acids must be removed or excluded. The precipitate then has the formula $(NH_4)_3PMo_{12}O_{40}.xH_2O$, and if it is dried at 200°C, cooled and weighed, it contains 1.65% of P, corresponding to $(NH_4)_3PMo_{12}O_{40}$. Floret also recommended boiling for half an hour after the oxygen-flask combustion, to ensure complete conversion into orthophosphoric acid. The Carius oxidation recommended here converts all the phosphorus into orthophosphoric acid without difficulty. Boiling for 1 min is sufficient.

15.1.2 Apparatus

Weighing tubes, type (b) as described in Section 3.1.
Conical flasks, Pyrex 100 ml.
Boiling-sticks. Pyrex tubes about 120 mm in length and 4.5 mm in diameter, fused near one end to form a cup (the lower end when in use). The other end can be sealed if desired. The cup should have no recess to trap precipitate. The sticks are used to prevent superheating during boiling of the solution, and must remain in the flask from the first boiling till the end of the determination. At least 6 boiling-sticks are required if a series of determinations is to be done.
Hot-plate. A small 3 heat electric hot-plate.
Small Pyrex 1 in. filter funnels. At least 6 will be required for series work.
Filter paper circles. Whatman No. 1, 5.5 cm.
Looped filter suction tubes. Pyrex tubes of about 6 mm outside diameter, looped near one end, and of total length 30 cm. They are attached to the stems of the filter funnels with small lengths of rubber tubing, loop end uppermost, and thus provide sufficient gravity suction to make the filtration conveniently rapid.
Burette, 50 ml. For quinoline molybdate solution. A burette with a coarse jet should be selected.

15.1.3 Reagents

Nitric acid, fuming (phosphate less than 0.002%).
Citric acid solution (10%). Make up about 50 ml (to a mark on a dropping bottle) on the day of use.
Quinoline molybdate solution (QM). Weigh into a 2 litre conical flask 150 g of molybdenum trioxide. Weigh into a 500 ml conical flask 30 g of sodium hydroxide pellets. Take 500 ml of pure water in a cylinder, and dissolve the sodium hydroxide in about 150 ml of it. Add the rest of the water to the molybdenum trioxide in the flask, followed by the sodium hydroxide solution. Mix and heat on the steam-bath for about 30 min. Filter while hot into a 3 litre flask. Add 460 ml of conc. hydrochloric acid and 2 drops of 100 vol hydrogen peroxide. This is solution M.

In a 2 litre flask put 300 ml of pure water, 300 ml of conc. hydrochloric acid and add to it, with mixing, 28 ml of freshly distilled quinoline. This is solution Q.

Add solution Q to solution M with shaking. Boil for 1 min (or keep on a steam-bath for 5 hr). Stand overnight, then filter into a 2 litre polythene bottle.

The solution is filtered into the burette used to dispense it. (For a recommended variation of the quinoline molybdate reagent see Campen and Sledsens [10]).
Phenolphthalein solution, 1% in alcohol.
Sodium hydroxide, 0.1*M*. A commercial standard solution, ordered in bulk, is

convenient. Store in polythene, and standardize against benzoic acid (120 mg ≡ approx. 10 ml) with phenolphthalein as the indicator.

Cellulose powder.

15.1.4 Procedure

Wash the weighing tubes with pure water, dry in the oven and allow to cool at least overnight under cover near the balance. Add a wisp (about 1 mg) of cotton wool to make a barrier to support the sample at the constriction. Put a 50 mg sodium chloride pellet in with the cotton wool. Weigh sufficient sample to contain ~2mg of phosphorus. A semimicro balance is usually adequate.

Wash the micro-Carius tubes with pure water, dry in the oven, cool and add 0.3 ml of fuming nitric acid from the safety pipette.

Put the sample tube into the Carius tube and seal it, if possible keeping the sample tube out of the acid until it is sealed. Place the tube in the cold heater, switch on, and leave for 3½ hr. The tube will then be at a temperature of about 300°C for about 3 hr.

After cooling (preferably overnight) open the tube (p. 30) and rinse into a 100 ml conical flask (giving a total volume of about 25 ml), allowing the weighing tube to drop into the flask. Add 1 ml of 2*M* hydrochloric acid. Put a boiling stick in the flask and boil for about 1 min or more on a hot-plate. Add 2 ml of 10% citric acid solution. Remove the weighing tube with a wire hook and rinse it in; add a pinch of cellulose powder. Keep the solution gently boiling on the hot-plate, and add slowly from the burette, with swirling of the flask, 10 ml of freshly filtered quinoline molybdate solution. If phosphorus is present, yellow quinoline phosphomolybdate will separate.

Boil for a few min longer, keep hot on the steam-bath for a total of about 15 min, then allow to cool before filtering. Decant the supernatant solution into a 5.5 cm filter paper in a 1 in. filter funnel fitted with a looped suction-tube. Wash the yellow precipitate with portions of about 5 ml of water by decantation, until the filtrate is neutral to blue litmus paper (9 or more washings). It is an advantage to tear off most of the two lower layers of the triple thickness of the folded filter paper and to ensure that the wet filter paper fits the funnel so snugly that it supports the column of water in the siphon tube. The filtration is then sufficiently rapid.

Return the filter paper and precipitate to the flask. Add about 10 ml of water and shake to disperse the precipitate. Slowly add 20 ml of 0.1*M* sodium hydroxide from a pipette (kept filled with the alkali when not in use). Shake until all the yellow precipitate has dissolved. (If this does not occur, the phosphorus content is greater than expected: add a further 20.0 ml of 0.1*M* sodium hydroxide.)

Add 5 drops of phenolphthalein solution. The solution should be pink.

Titrate with 0.1*M* hydrochloric acid until the pink colour just disappears. Standardize the acid against the alkali. Calculate the net volume of 0.01*M* alkali used.

1 ml of 0.1*M* alkali ≡ 0.1191 mg of phosphorus.

Example. The quinoline phosphomolybdate produced from 29.31 mg of a sample was dissolved in 20.00 ml of 0.1*M* sodium hydroxide. The back-titration was 2.88 ml of 0.1*M* hydrochloric acid; 19.80 ml of this acid were found to be equivalent to 20.00 ml of sodium hydroxide. Thus, the phosphorus content of the sample was

$$\frac{100 \times (20.00 - \frac{20.00}{19.80} \times 2.88) \times 0.1191}{29.31}$$

$= 6.95\%$ P.

15.1.5 OTHER METHODS

Numerous spectrophotometric methods have been developed, based either on the colour of phosphomolybdic acid or of the reduced form 'phosphomolybdenum blue'. A Kjeldahl decomposition is frequently used with these methods. A typical example is given by Chalmers and Thomson [11].

A method for determination of phosphorus and selenium is given by Binkowski and Rudnicki [12]. After decomposition of the sample with perchloric acid and nitric acid, the solution is neutralized, selenium is precipitated with thioacetamide and the phosphate is titrated with lanthanum.

15.2 THE DETERMINATION OF ARSENIC [13]

Arsenic can be determined in exactly the same way as phosphorus, except for the following slight modifications.

Take sufficient sample to contain about 4 mg of arsenic. After rinsing the Carius tube add about 0.1 g of potassium chlorate and dissolve it, then add 2 ml of 2*M* hydrochloric acid instead of 1 ml.

1 ml of 0.1*M* NaOH ≡ 0.2881 mg of As.

An arsenomolybdenum blue method can be used after oxygen-flask combustion of the sample [14] or an arsenomolybdic acid spectrophotometric method after a Kjeldahl decomposition [15]. A titration method for use after oxygen-flask combustion is given by Masson [16].

15.3 THE DETERMINATION OF GERMANIUM

A convenient method is based on spectrophotometric determination of germanomolybdic acid [15] after oxygen-flask combustion [17].

REFERENCES

[1] H. N. Wilson, *Analyst,* 1951, **76**, 65; 1954, **79**, 535.
[2] R. Belcher, *Submicro Methods of Organic Analysis,* Elsevier, Amsterdam, 1966, p. 76.
[3] S. C. J. Oliver, *Rec. Trav. Chim. Pays Bas,* 1940, **59**, 872.
[4] C. DiPietro, R. E. Kramer and W. A. Sassaman, *Anal. Chem.,* 1962, **34**, 586.
[5] R. Belcher and A. M. G. Macdonald, *Talanta,* 1958, **1**, 185.
[6] T. R. F. W. Fennell and J. R. Webb, *Talanta,* 1959, **2**, 105.
[7] H. Roth, *Pregl-Roth Quantitative Organische Mikroanalyse,* Springer, Vienna, 1958, p. 174.
[8] A. Floret, *Bull. Soc. Chim. France,* 1968, 1638.
[9] S. Posternak, *Bull. Soc. Chim. France,* 1920, 507; *Compt Rend.,* 1920, **170**, 930.
[10] W. A. C. Campen and A. M. J. Sledsens, *Analyst,* 1961, **86**, 467.
[11] R. A. Chalmers and D. A. Thomson, *Anal. Chim. Acta,* 1958, **18**, 575.
[12] J. Binkowski and A. Rudnicki, *Mikrochim. Acta,* 1977 **I**, 371.
[13] S. Meyer and O. G. Koch, *Z. Anal. Chem.,* 1957, **158**, 434.
[14] E. Celon, S. Degetto, G. Marangoni and L. Sindarelli, *Mikrochim. Acta,* 1976 **I**, 113.
[15] R. A. Chalmers and A. G. Sinclair, *Anal. Chim. Acta,* 1965, **33**, 384.
[16] M. R. Masson, *Mikrochim. Acta,* 1976 **I**, 399.
[17] M. R. Masson, *Mikrochim. Acta,* 1976 **I**, 385.

CHAPTER 16

Water by the Karl Fischer method

16.1 GENERAL DISCUSSION

Sufficient sample, if possible, to contain 0.5-4 mg of water is weighed into a titration vessel, which is immediately closed with a rubber serum cap to exclude atmospheric water. This titration vessel is equipped with a pair of platinum electrodes across which a potential difference of 80 mV is maintained. When iodine, from the Karl Fischer reagent, becomes present in excess, it depolarizes the electrodes and allows a current to flow. One ml of an anhydrous solvent is added to dissolve the sample. The sample is titrated directly with Karl Fischer reagent to the end-point, indicated by a sharp increase in current across the electrodes. There is a visible change from yellow to brown but the dead-stop method of end-point detection is far more sensitive and is essential if the sample is strongly coloured. For determination of the water content of solvents, up to 5 ml of sample can be titrated directly.

The water content is expressed as % w/w for liquid or solid samples, after due allowance for the solvent blank, and as % w/v for the water contents of solvents.

The principle of the Karl Fischer method is that sulphur dioxide will react with iodine to produce, in theory, sulphur trioxide and hydriodic acid only when water is present. Fischer [1,2] found that a solution of iodine in a mixture of anhydrous methanol and pyridine, to which an excess of sulphur dioxide was added, was suitable for titration of water, because the brown colour of iodine persisted only when all the water was consumed. A solvent containing -OH groups (methanol) and an organic base (pyridine) were both necessary, the latter because it formed salts with the hydriodic acid and sulphuric acid produced in the reaction with water.

Such a solution loses strength on standing, probably because of various side-reactions, but it must be remembered that unless moisture from the air is rigorously excluded, the reagent will lose strength by reaction with it.

A more stable reagent was devised by Peters and Jungnickel [3]; the methanol was replaced by 2-methoxyethanol (methyl cellosolve, which again contains an

-OH group). This stabilized reagent is obtainable commercially [4] and is recommended in the method described. Although Karl Fischer reagent should be restandardized when the the highest precision is required, this is not necessary so frequently with the stabilized reagent.

The analysis of an organic substance sometimes indicates that it is a hydrate. The easiest way to confirm this is by a Karl Fischer titration, provided the sample will dissolve in either pyridine or methanol. (Sometimes the sample dissolves more easily after some Karl Fischer reagent has been added.) Except when the water content is unusually high, it is advisable to take a fairly large sample. For example a monohydrate of molecular weight about 200 would contain about 9% of water, so 20 mg of sample would contain about 2 mg of water, which would require about 0.5 ml of Karl Fischer reagent. If the water content is lower, more sample should be taken, if it can be spared. Otherwise great care must be taken to check the blanks on the solvents as well as to make sure that the end-point is definite and stable and that moisture from the air is rigorously excluded.

The method described here is based on the micro-titration cell described by Levy *et al.* [5], and tested and developed by Corliss and Buckles [6]. The dead-stop end-point detector was suggested by Foulk and Bawden [7].

Although the Karl Fischer method has almost universal application, certain classes of compound can be analysed only with a modified procedure, and a few interfere totally with the reagent. Kolthoff and Belcher [8] give a comprehensive list of the classes of compound that require a modified procedure and those that cannot be analysed with the Karl Fischer reagent.

Although many hydrated salts of organic acids such as sodium tartrate have been suggested for the calibration of Karl Fischer reagent, in practice nothing has been found better than a weighed amount of water.

16.2 METHOD

16.2.1 Apparatus

Burette. A 2 ml burette with automatic pressure-filling from a 500 ml reservoir is modified so that it can be equipped with drying tubes to protect the reagent from the atmosphere and with taps to separate the silica gel in the drying tubes from the reagent when the apparatus is not actually in use. Alternatively, if the reservoir of reagent can be kept above the burette, the latter may be filled by gravity: again there must be provision for protecting the reagent and burette from the moisture of the air, and for a barrier, such as glass taps, between the Karl Fischer reagent and the drying tubes when the apparatus is being stored and is not in use.

Magnetic stirrer. A small variable-speed magnetic stirrer.

Dead-stop end-point indicator unit. A mains-powered indicator unit which may be assembled in the laboratory workshop or obtained commercially. The leads are fitted with small crocodile clips to make electrical connection with the platinum electrodes of the titration vessel.

Titration vessels. The titration vessels are made of Pyrex with platinum electrodes sealed through each side, diametrically opposite, with the ends reaching to the base of the vessel and inclined tangentially in order to provide room for a stirrer bar. They can be made by laboratory glassblowers from 13 mm bore 1 mm wall Pyrex tubing. Where the electrodes pass through the vessel the outside portion is twisted double so as to project for about 8 mm and epoxy cement (e.g. 'Araldite') is used to give strength and a liquid-tight seal at the junction with the glass wall of the vessel. A stock of at least 6 is kept. After drying overnight in the oven, the titration vessels are stored, ready for use, in a P_2O_5 desiccator.

Serum caps. Vaccine bottle caps, rubber, of such a size as to give an air-tight fit on the titration vessels. They are stored, ready for use, in a P_2O_5 desiccator. After some use these serum caps acquire a central hole and this can be used to accommodate the jet of the Karl Fischer burette.

Stirrer bars. Glass stirrer bars 1.2 mm in diameter and 8-10 mm long, to suit the titration vessels, are made by cutting a 5-7 mm length from an ordinary office staple and sealing it with a flame into a section of melting-point tube. These are kept ready for use in a P_2O_5 desiccator.

Syringe needle. A No. 15 (23 G × 1) disposable syringe needle is suitable as an air-bleed through the serum cap.

Desiccator (P_2O_5). A 4 in. desiccator containing phosphorus pentoxide.

Weighing sticks, as described in Chapter 2. About 10 are kept ready for use, resting in a 100 ml beaker and covered by a 400 ml beaker. After use, they are cleaned by washing with a suitable solvent, rinsed with acetone and allowed to dry at room temperature (to avoid an electrostatic charge).

Pasteur pipettes. Marked at 1 ml, and stored in an oven at 100°C until required for use.

Rubber teats. Used with the Pasteur pipettes and stored in a P_2O_5 desiccator until required for use.

Hypodermic syringe. A 1 ml syringe with a 2 in. needle may be used to measure samples for trace water determination. Store in a P_2O_5 desiccator until required for use. It may also be used to dispense the dried solvent, usually pyridine.

16.2.2 Reagents

Karl Fischer reagent, stabilized. A commercial preparation is normally used. Each ml is approximately equivalent to 4 mg of water. The burette reservoir is filled with one bottle at a time, and a spare bottle is kept in a refrigerator.

Methanol, anhydrous. This is available commercially as a special grade for Karl Fischer determinations. The water content is specified as less than 0.01% w/w (i.e. 1 ml is equivalent to less than 0.02 ml of Karl Fischer reagent). The bottle is kept well stoppered and a little molecular sieve (4A) is added to it.

Pyridine, anhydrous. Again, a special Karl Fischer grade is available commercially. This is kept well stoppered and about 5 g of molecular sieve (4A) is added to it. One ml is equivalent to about 0.02 ml of Karl Fischer reagent (less than 0.01% of H_2O).
Grease for the Karl Fischer burette. Rubber grease is often used. (Apiezon AP 101 grease and Sitco 300 grease have also been recommended.)

16.2.3 Procedure

Preparation. Rinse the serum caps with alcohol, wipe and store in the desiccator for at least 24 hr before use. Rinse titration vessels with water and dry at 100-150°C for at least 8 hr, transfer to the P_2O_5 desiccator and allow to cool before use. Rinse stirrer bars, dry between filter paper and store in the desiccator.

Fill the 2 ml burette with Karl Fischer reagent, using a rubber bulb to pressurize the reservoir. If the burette has not been used recently, rinse it out several times with the reagent beforehand.

Rinse the syringe and needle with acetone, blow dry with compressed air and store in the P_2O_5 desiccator.
Procedure for solid samples. Weigh out by difference enough sample (if sufficient is available) to contain 0.5-4 mg of water, into a titration vessel that has stood for at least 24 hr in a P_2O_5 desiccator. Add a stirrer bar and close with a serum cap from the desiccator.

Add 1 ml of a suitable anhydrous solvent to the sample, by using a Pasteur pipette direct from the oven and fitted with a dry teat. Alternatively, the dry solvent, usually pyridine, may be dispensed from the 1 ml syringe by passing the needle through the serum cap. Shake to ensure dissolution is as complete as possible.

Adjust the Karl Fischer burette to the 0.2 ml or other suitable mark, thereby filling the jet with fresh reagent, and immediately attach the titration vessel to the burette, by allowing the jet to pierce the serum cap (to exclude atmospheric moisture). Then insert a fine hypodermic needle into the serum cap to provide an air bleed. Connect the leads from the end-point indicator to the electrodes of the titration vessel and position the vessel centrally on the magnetic stirrer, ensuring efficient stirring. Add the titrant dropwise, allowing sufficient time between drops to ensure complete mixing. Before drops are completely mixed in, momentary deflections will show on the end-point indicator.

The end-point is reached when, with complete dissolution of the sample, the deflection on the meter is steady for 10-15 sec. Record the burette reading, V. (With colourless samples a change in colour from yellow to brownish can be seen.)
Calculation. If the initial reading of the burette is I ml, the final reading is V ml, the weight of sample is W mg, the blank for 1 ml of solvent is b ml, and the factor for the Karl Fischer reagent is F mg of H_2O per 100 ml of reagent, the water content is $[V - (b + I)]F/W$ % w/w of H_2O.

Procedure for viscous liquid samples. Take two titration vessels from the desiccator and, using the lighter of the two as a counterpoise, weigh out enough sample to give a titration of 1–2 ml. The sample may be transferred with a spatula or a Pasteur pipette, depending on its viscosity. Add a stirrer to both vessels and close both with serum caps.

Titrate the sample as for a solid. If it is too viscous, add 1 ml of anhydrous solvent to each vessel. Treat the empty counterpoise vessel in the same way and subtract the volume of reagent consumed, if any, from that required for the sample, to allow for any atmospheric moisture adsorbed by the vessel during the weighing.

Procedure for trace moisture in solvents. In this case plenty of sample is usually available, and not less than 5 ml should be made available to the analyst. The sample bottle should either have been thoroughly dried or well rinsed with sample before filling.

Place a stirrer bar in a titration vessel and fit it with a serum cap. By use of a 1 ml syringe and needle that have been kept in a P_2O_5 desiccator, inject 1 ml of sample, carefully measured, into the titration vessel. Titrate with Karl Fischer reagent as for a solid.

If the titration volume is small, add a further 1 or 2 ml of sample and continue the titration. Express the result as % w/v H_2O.

Determination of the factor of the Karl Fischer reagent. Take two titration vessels from the P_2O_5 desiccator and, using the lighter of the two as a counterpoise, quickly weigh out a single small drop of demineralized water (5–12 mg). Add a stirrer bar, cover with a serum cap immediately and titrate with Karl Fischer reagent. Do a duplicate determination.

Express the factor as F mg of $H_2O \equiv 100$ ml reagent. The duplicate results should agree to within 5 mg and the average value is then taken for the factor (initially about 400 mg $\equiv$ 100 ml). The factor must be redetermined if the reagent is more than 24 days old. Because the water is placed at the bottom of the vessel as a nearly spherical globule, a relatively small surface is exposed and a negligible amount of water is lost during the weighing. However, a large titration (2 or more burette fillings, i.e. 8 mg of water or more) is recommended for precise standardization of the Karl Fischer reagent.

Determination of solvent blank. Take a titration flask from the P_2O_5 desiccator, add the stirrer bar and close with a dry serum cap. Add 3 ml of solvent (methanol or pyridine) by using either a Pasteur pipette taken directly from the 100°C oven, or a syringe from the desiccator. Titrate as above. Express the result as ml of Karl Fischer reagent per ml of solvent. If this differs markedly from the previous determination repeat in order to confirm the result.

Anhydrous methanol and anhydrous pyridine should contain less than 0.01% w/v water. If the volume of Karl Fischer reagent required is greater than 0.1 ml per ml of solvent, the solvent should be discarded.

REFERENCES

[1] K. Fischer, *Angew. Chem.*, 1935, **48**, 394.
[2] J. Mitchell and D. M. Smith, *Aquametry*, 1st Ed., Interscience, New York, 1948; 2nd Ed., Part 3, 1980.
[3] E. D. Peters and J. L. Jungnickel, *Anal. Chem.*, 1955, **27**, 450.
[4] May and Baker, *Karl Fischer Reagent, Stabilized* (Brochure).
[5] G. B. Levy, J. J. Murtaugh and M. Rosenblatt, *Ind. Eng. Chem., Anal. Ed.*, 1945, **17**, 193.
[6] J. M. Corliss and M. F. Buckles, *Microchem. J.*, 1965, **10**, 218.
[7] C. W. Foulk and A. T. Bawden, *J. Am. Chem. Soc.*, 1926, **48**, 2045.
[8] I. M. Kolthoff and R. Belcher, *Volumetric Analysis*, Vol. III, Interscience, New York, 1957, Chapter 9.

CHAPTER 17

Loss of weight in vacuo

17.1 DISCUSSION OF THE METHOD

A stoppered glass tube containing a platinum boat is weighed against a counterpoise which consists of a similar glass tube with the stopper waxed in place and containing some copper foil instead of a platinum boat. The boat is removed from the glass tube and about half-filled with the sample (usually 100 mg), if this much is available. The boat is replaced in the tube, which is stoppered and weighed. The tube, boat and sample are then put into the drying apparatus, the stopper is removed and the whole is exposed to a high vacuum at a suitable temperature, in a space dried by phosphorus pentoxide.

The apparatus is allowed to cool under vacuum, then dry air is allowed to enter slowly, the stopper is replaced and the tube, boat and dried sample are again weighed. The loss in weight is thus determined. The drying must be repeated to constant weight. Sometimes the sample is volatile and the method is not suitable, or it may be unstable at the temperature used. A lower temperature should then be tried. Then it is usually better to dry at room temperature at least overnight.

Organic substances submitted for analysis frequently contain water or other solvents, sometimes in stoichiometric proportion (as in a hydrate).

Ideally, substances submitted for analysis should be anhydrous or free from the solvent from which they were crystallized. Sometimes, however, the anhydrous compound is hygroscopic and because of this a stable hydrate has to be submitted for analysis.

Often the water content can be determined rapidly by means of the Karl Fischer reagent but when this is not possible (for example because the substance is insoluble) the water, or other solvent present, may be determined by the loss in weight on heating in a vacuum.

The operation is simply that of weighing a sample before and after drying, but if the dried material is hygroscopic, it cannot then be weighed in an open vessel. Therefore, for all unknown samples, the method routinely adopted involves weighing the sample in a platinum boat in a stoppered glass tube before and after drying and using a device to close the stoppered tube inside the drying apparatus before removing it.

The apparatus and procedure described here has been used for many years for this purpose in the author's laboratory; it is very similar to that described by Clark [1] who followed the description given by Milner and Sherman [2].

Problems may arise, however, with polysaccharide materials, which can retain solvent up to temperatures at which the materials begin to decompose [3].

17.2 METHOD

17.2.1 Apparatus

The main glass part. This is shown in Fig. 17.1. It is advisable to keep several such pieces of apparatus complete and ready for use. A single retort clamp is used to hold the apparatus. This makes it easy to raise and lower it onto the heater. Wire springs are used where necessary to hold the various glass joints together. The springs are attached with wire (there is no need to have glass hooks made on the apparatus). All glass joints except L and M are lubricated with silicone high-vacuum grease. The drying tube C contains 'anhydrone' between cotton wool plugs.

Condensers. Several are kept set up ready for use so that several samples may be dried at the same time. The lower joint of the condenser is 19/26, as is joint L in Fig. 17.1.

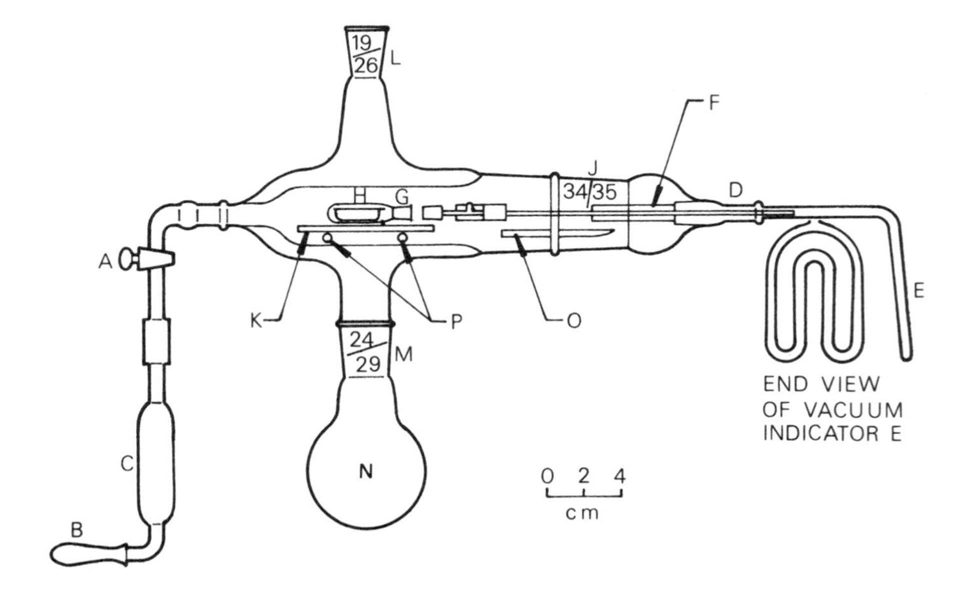

Fig. 17.1

Platinum boats. Old semimicro platinum boats may be used (~25 × 5 × 5 mm), or boats made to these dimensions from a 35 × 15 mm rectangle of 0.05 mm platinum foil in a similar way to the micro boats (p. 43). Although boats may be made from other metals (e.g. copper), platinum boats have the advantage that they may be very rapidly dried by ignition in a flame.

Glass weighing tube on aluminium plate. The weighing tube is made by modifying the weighing pig described in BS1428 [4]. The tail (handle) is removed and used to modify the handle of the stopper so that it is easily held in the brass bayonet holder (Figs. 17.2 and 17.3). A counterpoise is made from a similar

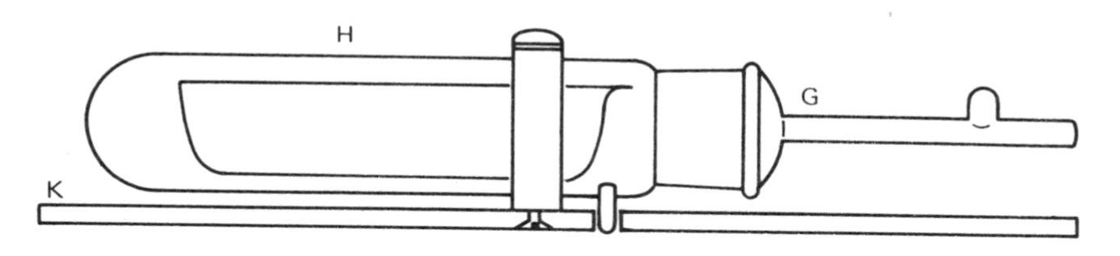

Fig. 17.2 – Platinum boat in weighing tube on aluminium plate

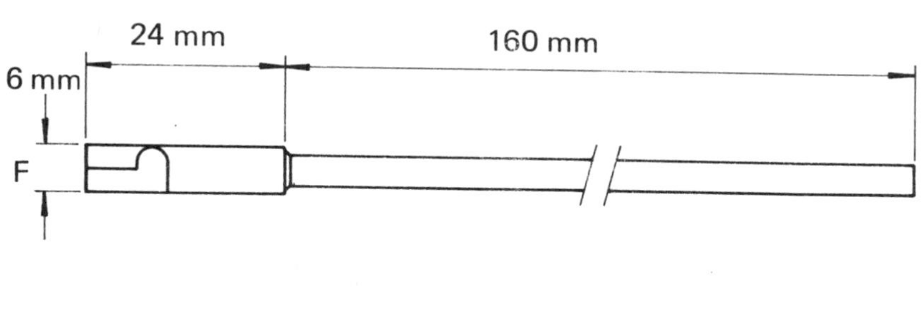

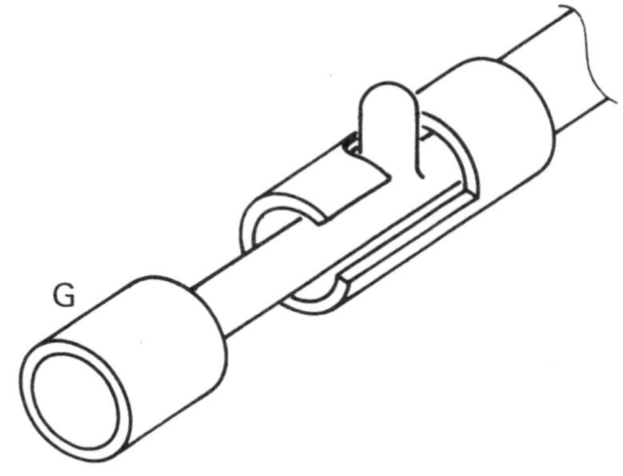

Fig. 17.3

tubes with its stopper waxed in place after the weight has been adjusted by putting copper foil inside it, trimmed to correspond to the weight of the platinum boat.

The one counterpoise is used for all the tubes and boats, but pieces of nichrome wire, made easily recognizable by their shapes, are provided as auxiliary counterpoises, one for each tube-and-boat combination.

Each glass tube is provided with an aluminium plate, about 70 × 25 × 2 mm with holes in it to accommodate the feet of the tube and (at one end) to engage with a wire hook. A Terry clip is screwed onto the plate so that the tube is gently gripped, but can be removed with tongs (outside the drying apparatus). It is then possible to remove the stopper by means of the brass bayonet device and the aluminium plate is prevented from rotating by the indentations in the inner tube of the drying apparatus (P, Fig. 17.1). The tube and platinum boat are weighed without the aluminium plate.

Troughs for phosphorus pentoxide. These are made by slitting Pyrex tubing lengthwise and are roughly 80 mm long, 22 mm wide and 8 mm deep (O, Fig. 17.1). Several spare troughs are kept in the drying oven, ready for use.

Wire hook. A simple hook of nichrome wire is provided for withdrawing the aluminium plate carrying the weighing tube. It engages in a hole in the end of the plate.

Chamois-tipped tongs. The tips of a pair of crucible tongs are covered by sewing on pieces of chamois leather.

Flasks for solvents. Several Pyrex flasks, complete with stoppers and cork rings, are kept for use, each with a particular solvent. They are round-bottomed 100 ml, with a 24/29 joint. They are also provided with wire round the neck so that a spring can be used to retain them on the main apparatus.

Electric bunsen burners. These are obtainable commercially (e.g. Electrothermal Engineering Ltd. Electrothermal Bunsen Series BA). They are a convenient form of heater for this purpose. The upper cover is removed and a 'Simmerstat' energy control is fitted. In use, the flask rests in contact with the heater and a very low setting of the Simmerstat is all that is needed to keep the solvent gently refluxing. Alternatively a heating mantle can be used.

17.2.2 Reagents

Phosphorus pentoxide.

Solvents for heating jackets. Chloroform (b.p. 61°C), n-butanol (b.p. 116°C), chlorobenzene (b.p. 132°C). Inflammable solvents are avoided if possible. Do not use steam as jacket vapour, if it can be avoided, because it causes clouding of the glass surfaces if used after an organic solvent. Butanol can usually be used instead, as its b.p. is quite close to that of water.

17.2.3 Procedure

Adjust the balance zero and put the counterpoise on the right-hand pan (a semimicro balance is adequate).

Bring the drying apparatus (without condenser) into the balance room. (After its previous use it will have been left evacuated and ready for use as described below.) Release the vacuum gently by turning tap A slowly, after having removed cap B from drying tube C. Release small joint D and remove the vacuum indicator E, placing it on the retort stand base so that the greased joint is not touching anything. By manipulating the brass tube F, place the stopper, G, in the glass tube, H, and detach the brass tube with its bayonet-fitting end. Loosen the large (34/35) joint J. (If it has been out of use for some time this may be difficult. If necessary try heating the joint gently in a flame.) Remove the joint with the brass tube inside it.

With the wire hook pull out the metal plate K with the glass tube and boat, and place it on the bench, touching with the fingers only the metal plate (which will not be weighed). With chamois-tipped tongs place the glass tube on the left-hand balance pan and weigh. (Weight of empty vessel.)

With chamois-tipped tongs place the glass tube on the notebook, remove the stopper with tweezers, remove the platinum boat with the wire and add sufficient sample, if enough is available, to half fill the boat (if less is available, even down to 10 mg, the determination is still often worthwhile). Replace the boat in the tube, stopper and reweigh (weight of vessel and sample).

Replace the glass tube in the clip on the metal plate and replace the whole in the drying apparatus, sliding it in as far as it will go, so that the metal plate is held by the indentations in the glass (P). Replace the trough containing phosphorus pentoxide. (When preparing for the second drying of a sample take a clean trough from the oven and put fresh P_2O_5 into it. Pass the spatula along the top of the trough so that the P_2O_5 does not stand up above the level of the edges of the trough. This reduces the risk of touching the P_2O_5 with the brass tube.) Reconnect the large joint J.

Remove the stopper from the glass tube by manipulating the brass tube with its bayonet-fitting end. Leave the stopper in the bayonet device and resting on the metal plate.

Replace the pressure indicator E. Take the entire apparatus to the high vacuum pump and evacuate for a minute or two. Close the glass tap A and replace the rubber cap B on the drying tube C. Attach a condenser (with cold water flowing) at joint L. Attach a flask containing a suitable solvent (usually butanol) to joint M. Lower the whole of the glassware by means of the one retort clamp so that flask N rests on the heater.

Adjust the control of the heater so that the solvent vapour surrounds the compartment containing the sample. No more heat should be used than to ensure that the upper joint L is hot to touch. After 30 min, or longer if convenient, raise the glassware and remove the heater. (During the 30 min, check from time to time, by observing the vacuum indicator E, that the vacuum is maintained.)

After 30 min for cooling, reweigh the glass tube, boat and sample as above. Before releasing the vacuum adjust the balance zero and replace the counterpoise and appropriate weights on the drying apparatus. The loss in weight, if any, can now be calculated as a percentage. The drying is repeated until constant weight is attained (i.e. the loss does not increase by more than 0.3 mg or 0.3% of the weight of the sample, whichever is the smaller).

If constant weight is difficult to attain, the sample may be volatile or undergoing thermal decompositon, and the determination should be repeated at room temperature. Volatility of the sample is sometimes indicated by the P_2O_5 becoming coloured (an organic sublimate being charred) or by sublimate forming on the brass tube.

If desired, the sample can be exposed to air (or in a hygrostat) and the gain in weight determined. Sometimes the sample will return to the original weight, a stable hydrate. The dried or equilibrated sample may be used for elemental analysis. More usually, the dried sample is no longer required for analytical work, but it should not be thrown away.

Empty the platinum boat, wash it with water and ignite it in a flame. Cool it on a metal block then replace in the glass tube. Place this in the drying apparatus as before and evacuate. The apparatus is kept in this condition so that it is always ready for immediate use.

REFERENCES

[1] E. P. Clark, *Semimicro Quantitative Organic Analysis,* Academic Press, New York, 1943.
[2] R. T. Milner and M. S. Sherman, *Ind. Eng. Chem., Anal. Ed.* 1936, **8**, 427.
[3] D. M. W. Anderson and N. J. King, *Talanta,* 1961, **8**, 497.
[4] BS1428: Part H1; 1960.

CHAPTER 18

Alkoxyl and *N*-alkyl

18.1 ALKOXYL GROUPS (the micro-Zeisel method)

When an organic compound which contains a methyl or an ethyl group attached to oxygen in its molecule is boiled with constant-boiling hydriodic acid solution, methyl or ethyl iodide is produced. The alkyl iodide is easily separated in a stream of carbon dioxide and collected, and its quantity is found by determination of the iodine contained in it.

To do this the iodine is oxidized to iodic acid by treating it with an excess of bromine. The excess of bromine is removed with formic acid and the iodic acid is determined by titration of the iodine liberated when it reacts with potassium iodide and dilute sulphuric acid.

18.1.1 Discussion of the Method

The micro-Zeisel alkoxyl determination has proved of great value over the years and a particular form of the apparatus was made the subject in 1954 of a British Standard [1], revised in 1962 [2].

In recent years it has become less needed partly because similar information is often obtained by nuclear magnetic resonance and partly because the simultaneous determination of carbon, hydrogen and nitrogen on modern instruments leaves little need for further analysis to identify substances made by synthesis.

However, occasions do arise when this method, so simple to carry out, proves of great value. It can be used, for example, to demonstrate the presence of methyl or ethyl alcohol when it is postulated that it is present as solvent of crystallization.

Colson [3] drew attention to the possibility, although rare, of reactive substances giving off the alcohol before it is converted into its iodide, thus causing low results, and he proposed a special apparatus to deal with such cases. This difficulty can also be overcome by heating the sample in a sealed glass tube with hydriodic acid, but neither of these devices is often required.

The method can be extended to n-propoxyl and isopropoxyl groups by heating in the same apparatus for a longer time with carbon dioxide flowing. Attempts to determine butoxyl groups usually give unreliable results.

The abundance of variations of the micro-Zeisel method that have appeared in the literature indicates its importance. One point of discussion is the nature of the scrubber (the purpose of which is to prevent iodine reaching the receiver in any form other than that of the alkyl iodide). The sodium acetate solution recommended is perfectly adequate, and even water is sufficient, if the manner of heating (to give a gentle reflux of hydriodic acid) and the rate of carbon dioxide flow are carefully controlled. However, to ensure that the sodium acetate solution in the scrubber is not alkaline, which could cause low results, it should be made very slightly acid with acetic acid. For a critical examination, and for the use of an infrared method, see Anderson and Duncan [4]. The various scrubber solutions have been critically examined by Belcher *et al.* [5] and by Bethge and Carlson [6].

The publication of the studies by Easterbrook and Hamilton [7] and by Heron *et al.* [8] led to the preparation of the British Standards for the apparatus [1,2] and the virtual standardization of the procedure for the micro-method. The later (1962) BS provided for the removal of bromine vapours from above the receiver solution by use of a water pump. This is an added complication to the apparatus and not considered justified by some.

18.1.2 Apparatus

The main glass part of the apparatus is shown in Fig. 18.1. It includes reaction vessel and condenser combined, which connects by means of a B14 joint to the scrubber, in which the carbon dioxide passes through a 7-turn glass spiral containing the washing solution, the scrubber being connected to the receiver by means of a B7 glass joint. The filter tube of the scrubber is closed in use with a piece of rubber tubing and a small glass rod. The receiver consists of a delivery tube, a glass rod spiral of about 17 turns, and a receiver tube, which has a bulbous enlargement (diameter about 28 mm) above the spiral.

A cylinder of carbon dioxide, analytical grade. A 7 lb cylinder, about 300 cm in overall height, is convenient. It should be fitted with a gas-pressure regulator. (A cylinder about 80 cm in height is also available and may be used.)

Bubble-counter and pressure-release valve. This is a combined piece of glassware connected by rubber tubing to the CO_2 needle valve and includes a safety device which will release any accidentally high pressure through a tube containing mercury. It also contains a bubble-counter which contains 50% sulphuric acid to show the rate of flow of the CO_2. Devices are included to prevent suck-back or carry-over of the acid. The bubble counter ends in a 2 mm glass tap. The exit tube of the glass tap is drawn out to an external diameter of 5-7 mm so that it can be joined to the 5-7 mm inlet tube of the reaction vessel with rubber tubing for easy connection and disconnection.

Dropping bottles, 100 ml, are used for formic acid, Methyl Red solution, iodine indicator solution, acidified sodium acetate solution, and hypophosphorous acid.

A burette with pressure-filling device, 10 ml, graduated at 0.02 ml intervals, is used for the 0.05*M* sodium thiosulphate (this is also used for iodine and bromine determinations).

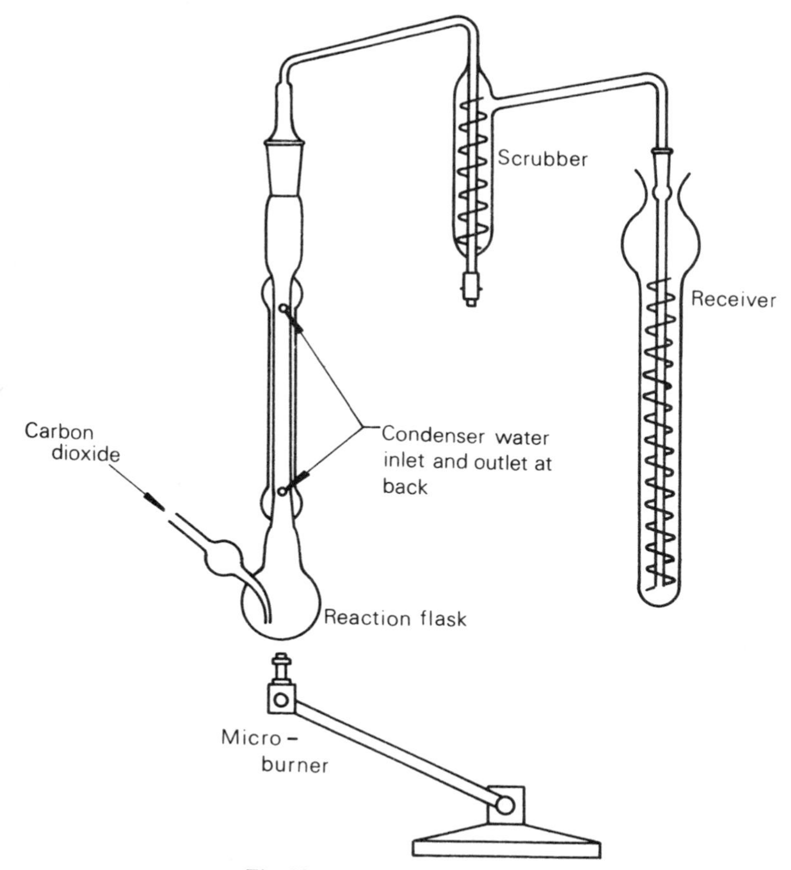

Fig. 18.1 – Micro-Zeisel apparatus

The titration flask is 100 ml, conical, made of Pyrex.
Glass weighing spoons (Fig. 18.2). These are described in BS 1428 [2] and are

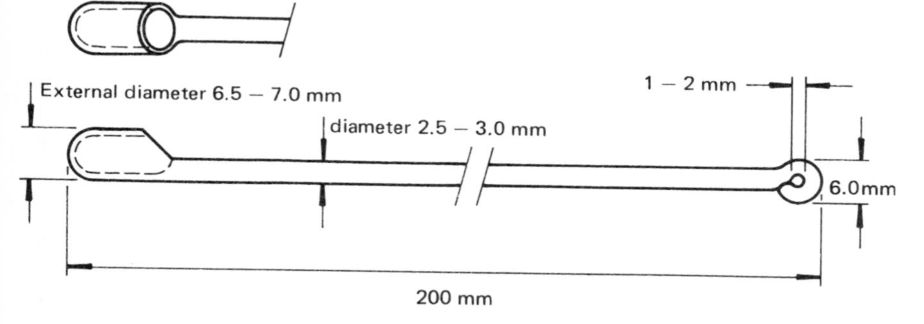

Fig. 18.2 – Glass weighing spoon for micro-Zeisel apparatus

made in soda-lime glass. About 10 of them are adjusted in weight to be a little heavier than a glass counterpoise of easily recognized shape which is kept in the balance case.

Micro gas burner. As described in BS1428 [9], modified for natural gas. It can be adjusted so that the jet is 25–152 mm above the bench.

Supports. Permanent bench scaffolding is used to support the scrubber by means of a small adjustable clamp. When the reaction vessel is not attached to the scrubber joint with steel springs it is loosely supported by the condenser leads and a loop of stiff wire attached to the upright of the bench scaffolding. The receiver is held in another adjustable clamp on the same upright.

A filling funnel made of glass, length about 55 mm, diameter at small end 10 mm and diameter at large end about 20 mm, for introducing reagents into the reaction vessel without any of them getting onto the B14 joint.

A 5 ml safety pipette.

18.1.3 Reagents

Phenol.

Hydriodic acid, 55% w/w HI, s.g. 1.7, in 5 ml sealed ampoules.

Solution of bromine and sodium acetate in acetic acid. Dissolve 200 g of $CH_3COONa.3H_2O$ in 100 ml of water by heating on a steam-bath. Cool, and dilute to 2 litres with glacial acetic acid. Take 500 ml of this solution and add 5 ml of bromine to it **(safety pipette)**.

Formic acid 90% w/w.

Methyl Red solution, 0.25% in ethanol.

Potassium iodide.

Sulphuric acid, 2N.

Sodium thiosulphate, 0.05*M*, as used for the determination of bromine (p. 94).

Iodine indicator solution, as for the determination of bromine (p. 94).

Hypophosphorous acid (30%).

Acidified sodium acetate solution (25%). Add about 5 g of sodium acetate trihydrate to a dropping bottle, dissolve it in water and make up to a 20 ml mark on the bottle. Make it very slightly acidic by dipping a thin glass rod into glacial acetic acid and then into the solution.

Silicone high-vacuum grease.

18.1.4 Procedure

Take a clean, dry reaction vessel (the part which includes the condenser). To clean a vessel which has been used 10 times, empty it, rinse with tap water and then with demineralized water, shake, and dry in the oven at about 100°C.

While the vessel is still hot put into it 2.5 g of phenol, roughly weighed, through a funnel placed in the neck (see below). The phenol will melt and run through the condenser and into the bulb. Connect the inlet tube to the bubble counter through which carbon dioxide is already bubbling.

Add 5 ml of hydriodic acid by opening the top of the ampoule and inverting it in the funnel, which is placed in the neck of the reaction vessel in order to avoid reagents, especially hydriodic acid, getting on to the ground-glass joint. Then, with a very hot glass rod heated with the blow lamp, crack the base of the ampoule, to let in air. This is the easiest way to transfer all of the 5 ml from the ampoule.

Attach the water leads and pass water gently through the condenser jacket.

Rinse the scrubber part of the apparatus with pure water, shake well and place in the glass spiral compartment about 2 ml of the 25% acidified sodium acetate solution (2 squeezes of a 1 ml dropper). Dry the B14 cone of the scrubber with clean paper and grease it with high-vacuum silicone grease.

Attach the reaction vessel to the scrubber with springs. The carbon dioxide will now bubble through the glass spiral in the scrubber.

Rinse a receiving vessel with pure water and shake it to remove most of the water. Put into it 15 ml (to a mark on the receiving vessel) of the bromine solution. Rinse with pure water the delivery tube carrying the glass spiral and put it into the receiving vessel, then attach the delivery tube to the B7 joint of the scrubber. The joint is moistened with water, well worked into place and secured with a tight rubber band. The carbon dioxide is now bubbling through the receiving vessel and each bubble as it rises should follow the long path within the spiral.

Weigh the sample, if a solid or non-volatile liquid, in a weighing spoon which has been rinsed with water and then with acetone and left to equilibrate in the balance room. Weigh the spoon empty and lying on the hooks of the balance pan. Then put the sample in the spoon (holding it with chamois-tipped tongs) and replace it on the balance hooks. Take, if possible, enough sample to give a titration of about 8 ml of 0.05*M* thiosulphate (this would be 9 mg of narcotine, for example, containing 22.5% of CH_3O–).

Disconnect the reaction vessel and support it by the loop of wire and the condenser leads.

Carry the weighing spoon and sample from the balance to the reaction vessel and insert it, sample end first, into the reaction vessel. Immediately reconnect the vessel, working the joint firmly together with the springs attached under high tension. If the glass hooks are not sufficiently far apart to cause high tension in the springs then the springs can be crossed or added lower down by bending stiff nichrome wire round the condenser.

Place the small flame under the reaction vessel, shake the vessel while the temperature is rising and adjust the flame to maintain a gentle boil for 1 hr for methoxyl compounds and $1\frac{1}{2}$ hr for ethoxyl compounds.

During this time ethyl or methyl iodide is removed from the boiling reaction mixture by the stream of carbon dioxide. The scrubber containing sodium acetate solution retains any iodine or hydrogen iodide, which would otherwise cause high results. The alkyl iodide is absorbed in the receiver where it is oxidized by bromine to iodic acid (HIO_3).

At the end of the appropriate boiling time, remove the receiver by first detaching the delivery tube at the B7 joint, and transfer its contents to a 100 ml conical flask by using a total of about 40 ml of water from a wash bottle for thorough rinsing.

Add about 8 drops of formic acid and shake. If the excess of bromine is not completely decolourized in 1–2 min, add further drops of formic acid one at a time. When the solution appears colourless add the smallest possible amount of Methyl Red solution. If the Methyl Red is bleached, bromine is still present. Continue adding formic acid and Methyl Red in drops, with pauses, until the colour of Methyl Red persists.

Add about 1 g of potassium iodide and 5 ml of 2*N* sulphuric acid. Titrate the liberated iodine with 0.05*M* sodium thiosulphate, using iodine indicator to show the last traces of iodine.

Each molecule of alkyl iodide is equivalent to one methoxyl group, formula weight of which is 31.03. Because the 'Leipert' amplification method is used, 6 atoms of iodine are produced from each molecule of methyl iodide. Hence 1 ml of 0.05*M* thiosulphate is equivalent to $31.034/6 \times 20 = 0.2586$ mg of methoxyl group or $45.06/6 \times 20 = 0.3755$ mg of ethoxyl group.

If both types are present together the alkoxyl oxygen can be calculated and 1 ml of 0.05*N* thiosulphate $\equiv 16.00/6 \times 20 = 0.1333$ mg of alkoxyl oxygen.

A blank is subtracted from the titration volume. This is usually about 0.1 ml of thiosulphate solution; it is determined by doing the determination with standard compounds two or more times.

18.2 *N*-METHYL (AND *N*-ETHYL) GROUPS

The sample is dissolved in phenol and propionic anhydride, and heated with constant-boiling hydriodic acid. Ammonium iodide and gold chloride are also present. The hydriodic acid is distilled off, but is collected in such a way that it can later be returned to the reaction-bulb. The dry residue, which consists mainly of the iodide of the sample, is then heated to about 350°C, at which temperature the *N*-alkyl group is split off as its iodide, which is estimated by determining the iodine collected in the receiver. To ensure that the reaction is complete the hydriodic acid distilled is returned to the reaction bulb and the process is repeated until no more alkyl iodide is produced. It has been found that 3 distillations are sufficient and that it serves no useful purpose to titrate the contents of the receiver at the end of each.

18.2.1 Discussion of the Method

Although many variations of the apparatus for this determination have been described, the committee concerned decided to describe the Friedrich apparatus [10] in BS1428: Part C1 in 1954 [1] and later in 1962 [2]. The author has used it as described here and has obtained good and useful results in many cases.

The procedure usually described is very tedious; the modified procedure given here is a more practicable method.

The drug hexamethonium, which contains two trimethylammonium groups, provides an excellent standard substance.

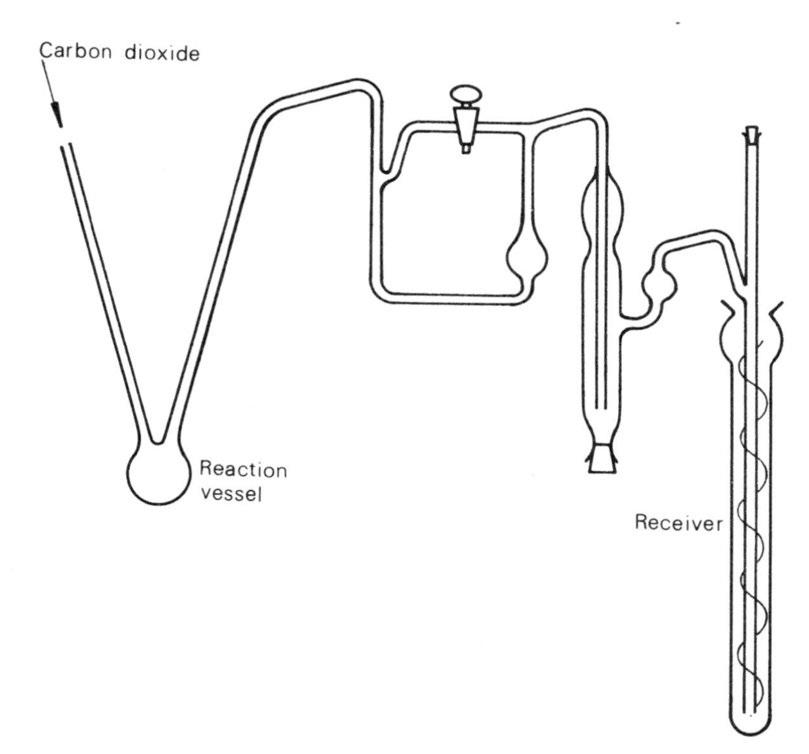

Fig. 18.3 – *N*-Methyl apparatus (Friedrich)

18.2.2 Apparatus

The main glass part (Fig. 18.3) is described in BS1428 [2].

A source of carbon dioxide gas is required (p. 152).

A bath of Wood's metal in a steel cylindrical container of about 90 mm diameter, capable of being easily raised and lowered. The bath can be heated easily on an 'electric bunsen' which also has a diameter of about 90 mm and can be supported on a laboratory jack.

A hot-water bath consisting of a 100 mm crystallizing dish. This can be heated on a 150 mm hot plate.

A glass weighing stick sufficiently narrow to go into the reaction bulb, with a counterpoise.

18.2.3 Reagents

The following are required in addition to those listed for the alkoxyl (micro-Zeisel) determination (p. 154).

Ammonium iodide.

Gold chloride solution. Dissolve about 0.1 g of gold in aqua regia. Evaporate to dryness on a steam-bath. Dissolve in about 5 ml of conc. hydrochloric acid and the minimum of conc. nitric acid.

Phosphorus suspension (red phosphorus, ~20%) in water.

18.2.4 Procedure

Clean the apparatus after the previous determination. This may be done by rinsing with hot water and leaving the reaction-bulb filled with and immersed in chromic acid, until next required. If rapid cleaning is required, conc. sulphuric acid may be boiled in the bulb (with addition of Kjeldahl catalyst) or hot concentrated potassium hydroxide solution may be used.

Put a 1 litre beaker of water on to boil and put the metal bath in the oven or over a small flame to melt the Wood's metal. Using the water-pump, suck boiling water through the apparatus, followed by cold water and then acetone. Dry by drawing air through and warming slightly with a bunsen flame, especially the reaction-bulb.

Put a little phosphorus suspension in the scrubber, only just sufficient to cover the tip of the jet (a small bulb pipette is kept ready for this purpose).

Weigh out 10–20 mg of sample into the reaction bulb. Use the special weighing stick for solids; weigh liquids into small glass cups which are dropped into the reaction-bulb.

Add about 0.1 g of phenol and warm the bulb in hot water to dissolve the substance. Add exactly 10 drops of propionic anhydride from the same pipette, kept ready for use. The substance must now be completely dissolved with warming in hot water if necessary. Add about 0.1 g of ammonium iodide and 1 drop of gold solution. Add about 3 ml of constant-boiling hydriodic acid from a 5 ml ampoule. Use a 3 ml pipette fitted with a rubber filling-bulb. Add one drop of hypophosphorous acid.

In the receiver put 13 ml of the solution of bromine in glacial acetic acid, then clamp the receiver in place. Put water in the capillary above the delivery tube and close it with a rubber bung. Put the close-fitting glass rod into the inlet tube and pass carbon dioxide by attaching the rubber tube from the cylinder. Adjust the rate of flow by means of the precision screw-clip so that the flow is fairly rapid and the receiver is full of rising bubbles. Clamp a 360°C thermometer as close as possible to the reaction-bulb so that the thermometer bulb is between the level of the liquid inside and the bottom of the bulb. Immerse the reaction-bulb in the metal bath with the metal surface level with the liquid in the flask. Immerse the continuous-loop receiver in nearly boiling water heated on a hot-plate.

Raise the temperature of the metal bath slowly to about 155°C, by using the energy regulator of the electric bunsen (about half full heat). The hydriodic acid will then distil and slowly collect in the first receiver. If hot hydriodic acid begins to enter the scrubber remove the heater and the metal bath, if necessary.

When the contents of the reaction-bulb are dry raise the metal bath so that the bulb is quite covered. Then raise the temperature to about 300°C. Allow the temperature to rise for a short while to 350°C but not above, and then bring it back to about 300°C. Keep for 30 min at 300°C or above. The carbon dioxide must be flowing as steadily as when initially adjusted. When the 30 min at 300°C has elapsed, remove both the water and metal baths and wait for 8 min for cooling to take place.

Return as much as possible of the distilled hydriodic acid to the reaction-bulb without sucking back the receiver solution. This is achieved as follows: (a) disconnect the carbon dioxide, (b) lower the receiver so that the delivery tube is just clear of the receiver solution, (c) turn the tap through 90°C, (d) apply mouth suction to the reaction vessel.

This will cause most although not all of the distilled hydriodic acid to suck back into the reaction-bulb. This transfer should be done quite rapidly (within a minute) and everything should then be replaced as it was, including the carbon dioxide flow and both heating baths in position as before.

Repeat the distillation of hydriodic acid and heating above 300°C as before, then suck back as before and repeat. The last heating should be at about 350°C for 30 min.

With carbon dioxide still flowing proceed to rinse out the receiver by raising the apparatus until the delivery tube is above the receiver solution. Remove the rubber bung and rinse the inside of the delivery tube with distilled water. Place the combined contents and rinsings of the receiver into a conical flask and complete the determination as for the alkoxyl method; that is, add formic acid until the bromine is destroyed (test with Methyl Red), add 2 g of potassium iodide and 10 ml of 2*N* sulphuric acid, and titrate the liberated iodine with 0.05*M* sodium thiosulphate.

Calculation

1 ml of 0.05*N* thiosulphate $= 15.03/6 \times 20 = 0.1253$ mg of methyl group or $29.06/6 \times 20 \equiv 0.2422$ mg of ethyl group.

REFERENCES

[1] BS1428: Part C1: 1954.
[2] BS1428: Part C1: 1962.
[3] A. F. Colson, *Analyst,* 1933, **58**, 594.
[4] D. M. W. Anderson and J. L. Duncan, *Talanta,* 1960, **7**, 70.
[5] R. Belcher, J. E. Fildes and A. J. Nutten, *Anal. Chim. Acta,* 1955, **13**, 16.
[6] P. O. Bethge and O. T. Carlson, *Anal. Chim. Acta,* 1956, **15**, 279.

[7] W. C. Easterbrook and J. B. Hamilton, *Analyst,* 1953, **78**, 551.
[8] A. E. Heron, R. H. Reed, H. E. Stagg and H. Watson, *Analyst,* 1954, **79**, 671.
[9] B.S. 1428: Part E4: 1962.
[10] A. Friedrich, *Mikrochemie,* 1929, **7**, 195.

CHAPTER 19

The equivalent weight of acids and bases

19.1 ORGANIC ACIDS

An amount of organic acid sufficient to react with just less than 10 ml of 0.01*M* sodium hydroxide (e.g. 10 mg of benzoic acid, m.w. = 122) is first dissolved in alcohol. Phenolphthalein is added and then 0.01*M* sodium hydroxide until the solution is distinctly pink. Hydrochloric acid (0.01*M*) is then added (approximately 0.5 ml more than is necessary to decolourize the solution), then the solution is boiled for about 30 sec. Provided it is still acid, the solution is then titrated, hot, with 0.01*M* NaOH until it remains just pink to the phenolphthalein for 1 min; otherwise more hydrochloric acid is added and the sequence of operations repeated. Corrections are applied for the acidity, if any, of the alcohol used and the net volume of alkali used is calculated.

Unreliable results may be obtained if the sample is not completely dissolved at the end of the titration.

19.1.1 Discussion

The method described is very simple although certain precautions have to be observed. Nowadays, a laboratory setting out to verify the identity of synthesized organic substances rarely needs to titrate acids or bases. Usually the determination of, for example, C, H, N and S will be adequate, especially when combined with infrared, nuclear magnetic resonance and mass-spectroscopic examination.

Consequently the more elaborate equipment needed for potentiometric or non-aqueous titration of acid and bases may not be justified, but where 0.01*M* acid and alkali are available (e.g. for Kjeldahl and boron determinations) the method described offers an easily followed procedure. It should be noted that if sodium hydroxide is consumed by an unknown sample it cannot be assumed that it contains a carboxyl group (–COOH). It may be an acidic phenol, such as picric acid, or a lactone or acid anhydride. The last two may usually be hydrolysed stoichiometrically to give the corresponding carboxylate if they are heated while alkaline, before completion of the titration to neutrality. Similarly, the equivalent weight will only be equal to the molecular weight when there is a single carboxyl group in the molecule.

As is well known, non-aqueous titrations make possible the titration of many organic substances which are too weak to be titrated in aqueous solution. The method is used extensively for quality control of drug formulations, and is often the basis of official assay methods. However, except where, for this reason or another, the reagents and apparatus are already to hand, the method is not really needed for synthetic organic compounds.

Potentiometric titration is also to be recommended if the equipment is already available and in constant use. Its advantage is that it will show a point of inflexion for each acid group, if more than one is present in the molecule, and will indicate the strength (i.e. p*K*-value).

The precautions needed for good results by the simple method described are as follows. First, the sample must finally be completely in solution. It often saves time, therefore, to start by making a solution in alcohol as described. Ethanol that has been freshly distilled from potash has been recommended. The propan-2-ol described in Chapter 8, which is dried and recovered by distillation from sodium hydroxide, can be used, if available. The removal and exclusion of carbon dioxide is also important. This can be assisted, and stirring effected as well, by bubbling purified nitrogen (or oxygen, see boron determination) through a side-tube into the titration vessel. The sodium hydroxide solution should be carbonate-free (ways of ensuring this have been described). (A solution of 0.005*M* barium hydroxide, in which barium carbonate would be insoluble, can be used.)

Quartz flasks have been recommended for this titration, but they are probably not necessary as modern borosilicate glass is usually satisfactory. Pure benzoic acid has been recommended as a standard. Another acid which is suitable, and which may also be used as a standard for fluorine, is *m*-trifluoromethylbenzoic acid. Pure potassium hydrogen iodate, $KH(IO_3)_2$, may be used to standardize the sodium hydroxide and also the sodium thiosulphate. Its very high molecular weight makes for high precision, but doubts have been expressed about the reliability of its purity [1]. Nevertheless, it is very convenient and an error of 0.1–0.2% in the purity is probably generally tolerable.

19.1.2 Apparatus

Burette for sodium hydroxide. Capacity 10 ml, graduated at 0.02 ml intervals. Its reservoir is a 500 ml polythene squeeze bottle which is also used for filling the burette (making a rubber bellows unnecessary) and requires only one guard-tube instead of the usual two. Air entering the burette passes over sodium hydroxide pellets in a 4 in. guard-tube. When the burette is not in use the guard-tube is separated from it by rubber tubing closed with a pinch clip, and is stoppered with a rubber bung.

Burette for hydrochloric acid. Capacity 10 ml, graduated at 0.05 ml intervals (also used for Kjeldahl nitrogen determinations).
Conical flask, 100 ml.
Glass cups. These are cut, by means of a scratch and a hot glass rod, from the bottom of specimen tubes (2 × ¾ in.), about ⅜ in. deep. (There is no need to flame-polish the cut edges.) After use, the cups are rinsed with distilled water and acetone and allowed to dry near the balance. Heat is not used.

19.1.3 Reagents

Alcohol. Ordinary laboratory 95% ethanol is used, but it is kept in a separate 500 ml bottle and labelled with its titre, e.g. 5 ml ≡ 0.14 ml of 0.01*M* NaOH, determined by titrating 5 ml of it plus 10 ml of water in exactly the same way as the sample.
Sodium hydroxide, 0.01*M*. Dilute 50.00 ml of commercial 0.1*M* solution to 500 ml in a standard flask, with demineralized water. This will fill the burette reservoir (450 ml) and allow some for rinsing. It is better to make up this solution frequently rather than to store a large quantity, which would continuously absorb carbon dioxide. Alternatively use the solution used for the boron determination (p. 187) which practically carbonate-free but may not be exactly 0.01*M*. To standardize, titrate 10.00 ml with 0.01*M* hydrochloric acid, using the procedure used for the samples. Then titrate a sample of pure benzoic acid as described in the procedure below, with correction for the titre of 5 ml of alcohol. Calculate the sodium hydroxide concentration from the weight of benzoic acid taken.
Hydrochloric acid, 0.01*M*, as used for Kjeldahl nitrogen determinations. Its exact concentration need not be known. To make, dilute 500 ml of the commercial 0.1*M* acid to 5 litres with demineralized water.

19.1.4 Procedure

Rinse a 100 ml conical flask with distilled water and alcohol and leave the flask wet with the alcohol. Weigh into the flask approximately a tenth or less of the expected equivalent weight in mg. Use a small glass cup which can be dropped into the flask and left there until the titration is completed. Unless the sample is known to be easily soluble in cold water, add 5 ml of alcohol and dissolve the sample; otherwise use 5 ml of water. For a solid, first add about only 4 ml and dissolve the sample with warming if necessary. Then add the rest of the alcohol, rinsing down the walls of the flask. Some solids are too insoluble to dissolve in the alcohol even when hot, but they may dissolve in water after an excess of sodium hydroxide has been added. This should be tried next. Add 3 drops of phenolphthalein solution. Titrate with sodium hydroxide until just pink. Add 0.01*M* hydrochloric acid from the burette, until the colour is discharged, then about 0.5 ml more. Add a small piece of porous pot and bring the contents of the flask to the boil over a micro-burner and keep boiling for about 30 sec.

Complete the titration of the now hot solution with the 0.01*M* alkali till a faint pink colour persists for 1 min. Record the volume of acid and alkali used.

Calculate the equivalent weight as in the following example.

Weight of organic acid taken = 9.372 mg
Volume of 0.01*M* NaOH used = 8.34 ml
Alcohol blank titre = 0.14 ml
Corrected volume of alkali = 8.20 ml
Volume of 0.01*M* HCl used = 0.53 ml
10.00 ml of the alkali requires 10.21 ml of the acid
Corrected volume of acid = 0.54 ml
Net volume of alkali used = 7.66 ml

In standardization, 12.21 mg of benzoic acid required a net volume of 12.26 ml of alkali, so the NaOH is 0.00994*M*.

Hence, equivalent weight = 9.372/7.66 × 0.00994 = 123.1. Or, as a formula, $E = w/v \times f$ where w is the weight of sample in mg, v is the volume in ml of NaOH required to neutralize it, f is the molarity of the NaOH, and E is the equivalent weight.

19.2 ORGANIC BASES

The organic base is titrated against 0.01*M* hydrochloric acid, to Methyl Red indicator (screened with Methylene Blue). This indicator is suitable because the organic base is likely to be weak compared with hydrochloric acid. An excess of acid is first added and the excess is titrated with 0.01*M* sodium hydroxide. Since the titration is carried out to a Methyl Red end-point the carbonate content of the small volume of alkali used may be neglected.

Weak bases often give indefinite end-points in the titration but this observation is often of value in characterization of the sample.

The strength of organic bases can vary enormously. Some may be so weak that indefinite results are obtained by the method described, but others may be strong bases that will absorb carbon dioxide from the air. They may then have to be weighed in a fragile sealed glass bulb which can be crushed in an excess of 0.01*M* acid.

19.2.1 Apparatus

As for titration of organic acids.

19.2.2 Reagents

In addition to those used for titration of organic acids, the following are required.

Methyl Red solution (0.5% in alcohol, as used for Kjeldahl nitrogen determination).
Methylene Blue solution (0.1% in water, as used for Kjeldahl nitrogen determination).

Ethanol, 95%. The same stock bottle is used as for organic acids, and the titre of 5 ml (plus 10 ml of water) should again be found in exactly the same way as for the titration of samples.

19.2.3 Procedure

Weigh out a sample as for organic acids. Add 1 drop of Methyl Red solution. Add $0.01M$ hydrochloric acid from the burette with swirling, until the solution is definitely red, and a further 0.5 ml.

Add 1 drop each of the Methylene Blue and Methyl Red solutions and titrate with $0.01M$ sodium hydroxide until the violet colour just disappears. A drop or two in excess should produce a green colour. (When the organic base being titrated is very weak this end-point may be rather indefinite, but helps to characterize the sample as a weak base.)

The result may be expressed either as an equivalent weight or as a percentage of nitrogen. The nitrogen found in this way may be a whole number fraction of the total nitrogen content.

Record the volumes used (x ml of $0.01M$ HCl, y ml of $0.01M$ NaOH). Subtract the titre of the alcohol from x when applicable, to give a corrected volume z ml. Convert y into the equivalent volume u of the $0.01M$ hydrochloric acid, after titrating one with the other (same indicator).

Calculate the equivalent weight $E = w/(z-u)f$ where w is the weight of sample in mg and where f is the molarity of the hydrochloric acid (for rough work both the acid and alkali may be assumed to be $0.0100M$ if the solutions are made from commercial $0.1M$ solutions).

Alternatively calculate the percentage of nitrogen in the sample as $(z-u)f \times 1401/w$.

REFERENCE

[1] Analytical Methods Committee, *Analyst*, 1965, **90**, 251.

CHAPTER 20

Acetyl and *C*-methyl

When an organic substance which contains an acetyl group attached to an oxygen or to a nitrogen atom is boiled with sulphuric or phosphoric acid of a suitable concentration, acetic acid is produced by hydrolysis. If the water is then distilled, more water is added and distilled, and this process repeated as often as is necessary, a quantitative yield of acetic acid is obtained in the receiver. This can be titrated with 0.01*M* sodium hydroxide with phenolphthalein as indicator. A compound containing a benzoyl group will give benzoic acid in the same way (also the salicyl group).

If an organic compound is boiled with chromic acid of suitable concentration it will be destroyed, except that methyl groups attached to a carbon atom (terminal or *C*-methyl groups) will survive as acetic acid, which may then be distilled and determined in a similar way. Certain types of *C*-methyl groups, in particular where two methyl groups are attached to the same carbon atom (*gem*-dimethyl groups) do not give quantitative yields of acetic acid, but use may be made of the fact that similar groupings usually give similar yields of acid.

The determination became very much easier to perform when Wiesenberger described his apparatus [1]. In a simplified form this was later made the subject of a British Standard [2].

20.1.1 Discussion of the method

The apparatus described has changed what used to be a very tedious procedure into a very practicable method. The acetyl (or benzoyl) determination sometimes fails through frothing, but it can then often be done as a *C*-methyl determination (because the acetyl group, CH_3CO-, contains a *C*-methyl group), if the ingredient causing frothing is destroyed by the chromic acid.

Hydrolysis with alcoholic sodium hydroxide has sometimes been recommended (followed, after acidification, by distillation from acid). This has very frequently been found to cause frothing (probably by producing substances resembling soaps). It is therefore advisable to try chromic acid in preference to this.

20.1.2 Apparatus

The main apparatus (Fig. 20.1) is as described in BS1428 [2], except that it is recommended that the ground-glass joint between the still-head and the condenser be eliminated and the two components joined by fusion, at the same time the distance between the still-head and the cold condenser surface being made as short as possible. (Wiesenberger recommended water-jacketed ground-glass joints, but although these were typical of his thoroughness, they are not really necessary.) In addition another *Liebig condenser* with B19/26 cone at the lower end is required for attachment to the reaction flask for the first hydrolysis stage.
Reaction flasks. The flasks of the shape shown in Fig. 20.1 and in the BS have a B19/26 socket. It is convenient to have 3 such flasks available: one in use, one rinsed and drying in the oven and one dried and cool ready for use. They can then be used in rotation with the minimum of delay. A glass pip is made on the wall of the flask to indicate the 4-ml level.
A 5 ml safety pipette (graduated at 0.05 ml intervals).
A 100 ml conical flask for the titration.

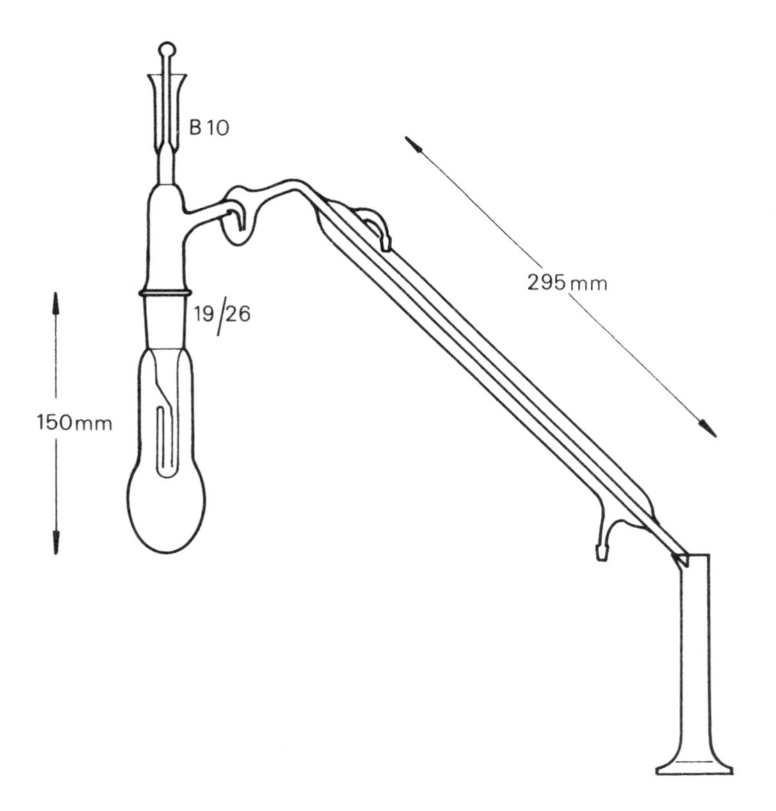

Fig. 20.1 – Acetyl and C-methyl distillation apparatus

A 25 ml measuring cylinder to be used as a receiver. The position of the condenser clamp, once adjusted to suit this receiver, does not need any further adjustment.
Anti-bumping granules.
Electric bunsens.

20.1.3 Reagents

Wenzel's sulphuric acid. Add 100 ml of conc. sulphuric acid to 200 ml of water with swirling and then boil it in the open flask for about 30 min, adding water from time to time to bring it back to its original volume. This boiling is intended to eliminate any steam-volatile acids.
Phosphoric acid. Treat 100 ml of phosphoric acid (s.g. 1.710) and 100 ml of water similarly to the sulphuric acid.
Chromic acid solution. Dissolve 67 g of chromic oxide in 500 ml of water and add 125 ml of conc. sulphuric acid with care and swirling. Boil for 30 min, replacing the water boiled off.
Silver sulphate.
Sodium hydroxide, 0.01*M*, as used for the boron determination (p. 187, made from washed sodium hydroxide pellets, boiled and cooled water, and always protected from the air by soda-asbestos guard-tubes). It may be standardized against pure benzoic or sulphamic acid.

20.1.4 Procedure

C-Methyl determination. The sample is completely oxidized by means of boiling chromic acid solution to yield carbon dioxide and water (and other inorganic products when elements other than carbon and hydrogen are present) except for any *C*-methyl groups which, ideally, yield the equivalent quantity of acetic acid. For certain types of compound, low but often consistent results are obtained. It should be noted that the *C*-methyl procedure differs from the acetyl procedure only in the timing and the reagent used. The *C*-methyl procedure is sometimes easier to apply than the acetyl determination, because there is usually less frothing in the distillation after the oxidation.
Acetyl and benzoyl group determination. The corresponding acid, acetic or benzoic, is liberated by boiling under reflux with sulphuric or phosphoric acid. Phosphoric acid is preferable if there is any risk of producing sulphurous acid which would be carried over into the receiver and cause high results. In both cases the acetic or benzoic acid is distilled out of the reaction mixture (together with relatively large volumes of water) and titrated with 0.01*M* sodium hydroxide, with phenolphthalein as indicator.

Weigh up to 100 mg of sample (enough to give a 10 ml titration, if possible) into a dry, cold reaction flask, by means of a long-handled weighing stick. If halogen is present, add about 20 mg of silver sulphate.

Rinse the reflux condenser with water. It is not necessary to dry it completely, but to avoid too great a dilution of the reagent partly dry it by pushing a small filter paper circle through it from top to bottom with a clean glass rod. This soaks up much of the adhering water in a consistent manner. Attach the flask with the sample in it to the reflux condenser. Make a perfect joint by moistening with water and attaching strong springs.

Add through the condenser, from the safety pipette, 4 ml of Wenzel's sulphuric acid for acetyl and benzoyl determinations, or 4 ml of chromic acid solution for *C*-methyl determination. (Use phosphoric acid for the acetyl determination instead of sulphuric acid if sulphurous acid might be formed.) Add an anti-bumping granule and reflux gently for 30 min for acetyl and benzoyl, and for 90 min for *C*-methyl determinations. Heat is supplied from an electric bunsen with the energy regulator at a predetermined setting.

After the specified time of gentle boiling, remove the heater and cool the flask by rinsing with cold water followed by immersion in iced water for 5 min. This is to avoid losses of the volatile acid, when the flask is opened during its transfer to the distillation apparatus. Rinse the condenser into the flask adequately but with the minimum of water. Add two more anti-bumping granules. Transfer the flask to the distillation apparatus and make a good joint by moistening it, working it well into position and attaching strong springs. (At this stage start the reflux stage of the next determination, if any.) Make sure cold water is flowing in the condenser jacket, and distil, using an electric bunsen as the source of heat. Add water up to the mark in the cup above the still-head. The distillation should be started gently at first, but once under way it should be as rapid as possible. This is not only to save time but also to minimize the amount of water collecting in the two traps so that they do not have to be emptied too frequently (done by removing the heater, and as soon as the traps are emptied by the cooling effect, replacing the heater so that bumping is not caused by letting the solution stop boiling). The object at this stage is to remove all the water that was used for rinsing so that the volume and acid concentration return to what they were for the hydrolysis stage (to the 4 ml mark on the flask). From then on, 2 ml of water are alternately added from the cup and distilled off again, until a total of 30 ml has been collected (not counting the rinsing water). As rapidly as possible the initial distillate in the 25 ml measuring cylinder is transferred to the 100 ml titration flask and the distillation started again. Then 20 ml are collected in the measuring cylinder (by the alternate addition and distillation), emptied into the titration flask, and a further 10 ml distilled. This has been found to be sufficient and not unduly time-consuming. The entire distillation can be done in about 20 min. If it takes longer, it is not being done in the proper way and the traps are filling too quickly. (Unfortunately with acetyl determinations the mixture sometimes froths and the distillation is then very slow or even impossible; in such cases a chromic-acid oxidation is often easier to use.) Titrate the combined distillates with the 0.01*M* sodium hydroxide, after adding 3 drops of phenolphthalein solution. Take the end-point when a very faint pink has persisted for 3 min (more easily seen if the solution is poured into the measuring cylinder and viewed from the top against a white background). This also serves to rinse the receiver and to give a final cautious approach to the end-point.

Rinse the reaction flask with water, allow it to drain, and dry it in an oven. The distillation apparatus has now of course been submitted to a thorough cleaning process so it can be used again without further cleaning, if kept clean by covering its ends. The receiver and titration flasks are stored inverted in wire racks and are thus ready for immediate use. Three reaction flasks are used in rotation. When the apparatus is not in use, one flask is attached to the distillation apparatus to keep it clean, the second is inverted in a wire holder, clean and dry ready for use and the last is in the oven drying. Thus the apparatus is always ready for immediate use and determinations are easily carried out in rotation.

Calculation

Acetyl = CH_3CO- = 43.04; 1 ml of 0.01*M* sodium hydroxide ≡ 0.4304 mg of acetyl group: *C*-methyl = CH_3- = 15.03; 1 ml of 0.01*M* NaOH ≡ 0.1503 mg of methyl group.

REFERENCES

[1] E. Wiesenberger, *Mikrochem. Mikrochim. Acta,* 1947, **33**, 51.
[2] BS1428: Part C2: 1954.

CHAPTER 21

Molecular weight

For the complete characterization of a new and previously unidentified organic compound a determination of molecular weight is usually necessary, but for a compound synthesized in the laboratory it is often not absolutely essential, because the molecular formula may not be in any doubt. If the method of synthesis is known, good guesses can often be made on the basis of the elemental analyses. For example if 3.2% sulphur has been found, this implies a molecular weight of 1000, or 2000 if 2 sulphur atoms are present. Often the multiple values are unlikely. If the compound has been extracted from natural sources, however, such as from plants or animals, a molecular weight determination is much more important, although if a mass spectrum is available – and it usually is for such work – then this usually provides strong evidence of the molecular weight when considered together with other evidence such as the elemental analysis.

Consequently traditional methods of determining molecular weights are seldom used now, but brief descriptions of two such methods follow. The Rast [1] method is based on the lowering of the melting point of a solid such as camphor when a substance is dissolved in it, and the ebullioscopic method is based on the raising of the boiling point of a solvent when a substance is dissolved in it.

Although other methods of molecular weight determination are known, particularly those based on vapour pressure, the author has confined himself to these two methods, which he has found to be the most useful in a laboratory mainly devoted to organic elemental analysis.

It is recommended that the ebullioscopic method should normally be tried first, once it has been shown by a simple test that the sample will dissolve to give a solution of about 1% concentration in boiling acetone, methanol or benzene, in that order.

It frequently happens that the sample is almost insoluble in these solvents (which is a hint that the molecular weight may be high). Then the Rast method may be tried.

21.1 THE RAST METHOD

This method, better called the cryoscopic method, depends on the fact that camphor (and other substances), when melted, can dissolve many organic substances, and also on its molal melting-point depression. If a mole of a substance is dissolved in 1000 g of camphor the melting point of the latter is lowered by 40°C. If water is the solvent the corresponding figure is only 1.86°C. As camphor is a volatile solid (molal melting-point depression is related to the vapour pressure and the latent heat of fusion of the solvent) it is necessary, to prevent loss of camphor by evaporation, to seal the solvent and solute, after they have been weighed out, into small glass tubes in which the solution can be made by heating and agitation. The ingenious devices of Pirsch [2] and others described below allow weighed amounts of solvent and solute to be sealed up without causing either to be heated, and perhaps decomposed, when the glass tube is closed in a flame. Special techniques are needed if the sample is a liquid, and the substance must not only be soluble in camphor but also be stable at the melting point of camphor (180°C). If these conditions are not fulfilled, there are many other solvents which can be tried, the most notable being camphene (m.p. 49°C) which has a molal melting-point depression of 31 °C, and dihydro-α-dicyclopentadienone (m.p. 53°C), molal melting-point depression 92°C [3] – more than twice the remarkably high figure for camphor. Some thirty suitable solvents have been described, usually hydroaromatic substances, but, once again, a laboratory devoted to elemental analysis can usually be satisfied with camphor and camphene and perhaps one or two others. Wider applications of the method are better left to specialists.

21.1.1 Apparatus

Melting-point tubes 2-3 mm in diameter, made from glass tubing. Suitable small glass tubes and rods are shown in Fig. 21.1. They are kept ready for use in a desiccator. The melting-point tubes should be placed in the balance some minutes before weighing.

A melting-point apparatus which consists of a heated bath of a clear liquid in which can be immersed an appropriate thermometer to which the melting-point tube is attached. For melting points below about 90°C the liquid may be water.

Thermometer. The thermometer most often required is one with a range from 140 to 190°C, graduated at 0.5°C intervals. Thermometers made especially for the purpose may be obtained. If a solvent other than camphor is used another suitable thermometer is desirable (graduated at 0.1°C intervals and covering the range from 10°C below the m.p. of the solvent to 40°C above it). Those known as Anschütz thermometers are suitable, although the absolute value of the melting-point is not required accurately. It is the magnitude of the melting-point depression that is important.

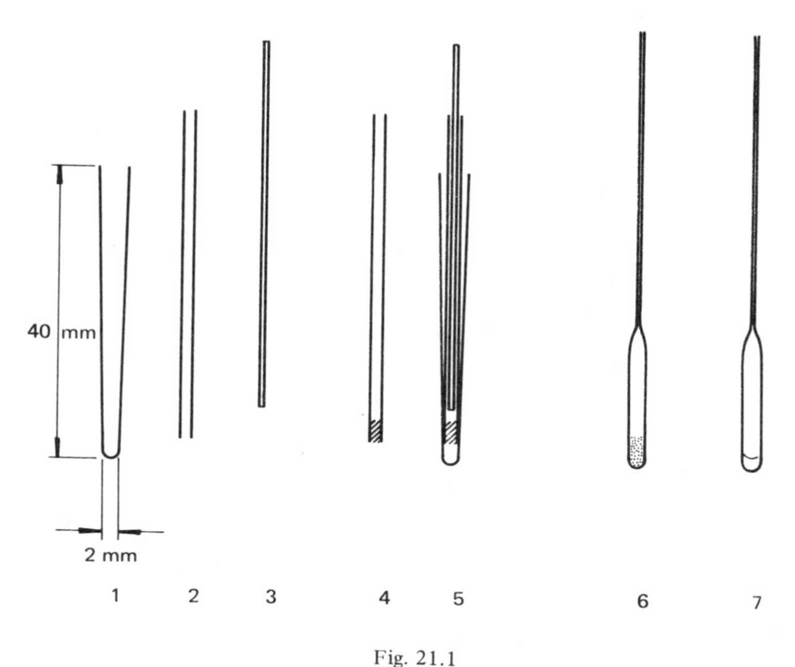

Fig. 21.1

21.1.2 Reagents

Camphor. Ordinary pure or BP grade camphor is suitable. It may be improved by breaking up the rather waxy pieces in a mortar after moistening with diethyl ether. The ether is then allowed to evaporate by leaving the solid spread out on filter paper. A granular and uniform camphor can be obtained in this way.
Camphene. This is obtainable from suppliers of microanalytical reagents.

For other solvents see Roth [4].

21.1.3 Procedures

Solid samples. Place one of the small tubes (1 in Fig. 21.1) on the balance pan and weigh it. Take a narrow glass tube (2) and pick up in its end about 0.2-1 mg of the sample (4). Clean the outside of this tube with a tissue. Put the rod (3) inside tube (2) and put the latter into the melting-point tube (5) so that the sample may be deposited at the bottom, by pushing the rod, and so that none of the sample adheres further up the walls of the tube (1). Then weigh the tube again to find the weight of the sample. Add camphor in a similar way (about 10-20 times the weight of the sample), using a second small tube and not touching the sample. Weigh again to determine the weight of camphor added.

Seal the tube in a small flame to produce a bulb 15-20 mm long, with a long glass handle about 45 mm long (6). Now dip the sealed tube in a bath of liquid, such as liquid paraffin, at a temperature just above the melting-point of camphor, 180°C. Agitate, and the whole should melt to give a clear solution (7). If some solid remains undissolved the experiment should be repeated with proportionately more camphor. If that fails, other solvents should be tried, because a clear solution must be obtained.

Prepare a similar sealed tube containing camphor only. Attach both tubes to the thermometer with a small rubber band so that the contents are level with the mercury bulb, and immerse in the heating bath. Raise the temperature of the bath, with stirring, at quite a rapid rate and observe at roughly what temperature melting occurs. This can be taken as the temperature at which the very last crystal of solid disappears. Then allow the bath to cool slowly. It may be found easier to take the temperature at which solidification begins, which usually appears as crystals dropping out of the clear solution like snow. In any case, the observations should be made several times and the same method must be used for the camphor alone as for the solution in camphor. The difference between the melting or freezing temperatures may be called 'ΔT'.

The experiment must also be carried out with a standard substance (naphthalene and benzoic acid are suitable) and the constant, K, for the particular sample of camphor determined.

$$\text{Then } K = \frac{M \times L \times \Delta T}{1000 \times S}$$

where M is the molecular weight of the standard, L is the weight of camphor in mg and S is the weight of sample in mg.

As K is about 40, and there should be about 10 times as much camphor as standard (benzoic acid, m.w. 122), then ΔT should be

$$\frac{40 \times 1000 \times 2}{122 \times 20} \sim 33°\text{C}.$$

The depression in this case would be about 33°C.

Clearly the molecular weight of an unknown sample will be given by

$$\frac{K \times 1000 \times S}{L \times \Delta T}$$

where K is the average found from several standardization experiments.

Once found, the figure for K can be used for as long as the same batch of camphor is used.

Liquid samples. When the sample is an oily, viscous and non-volatile liquid the method of Pirsch [2] may be used. Into a glass tube that is 80-90 mm long, open at both ends, and of outside diameter not more than 2.5 mm, a glass rod of about 0.8 mm diameter is fused exactly centrally (Fig. 21.2). The glass rod

Fig. 21.2

projects by about 1–1.5 mm at one end (the left in the diagram). When this end is dipped carefully into the liquid sample, a suitable amount usually remains attached to it. When this tube is then carefully lowered to the bottom of a weighed melting-point tube held vertically, sufficient of the liquid will usually be deposited. The outer glass tubing acts as a sheath preventing deposition of any of the liquid higher up the tube. The camphor (or other solid solvent) may be added in the same way as for solid samples. The tube is weighed and sealed and the determination is completed in the same way as for solid samples.

High-boiling liquids may be weighed out by the method of Soltys. Contamination of the walls of the melting-point tube with sample can be avoided, if the liquid sample is introduced with a small glass tube which is drawn out at its narrower end to a hair-fine capillary of length 1.5–2 mm. The capillary is dipped into the liquid and the outside is well wiped. It is then introduced into the bottom of a weighed melting-point tube, the sample is blown out, and the filling tube is carefully removed. This usually puts about 0.3 mg of the liquid into the tube without contaminating the upper walls. The camphor (or other solid solvent) can follow in the usual way and the determination is then completed as for a solid sample. By this means the liquid sample is usually so readily absorbed by the camphor that the melting or freezing point first determined does not change when repeated.

Liquids of low boiling point may be weighed out by Pirsch's method [2]. First, the solvent (e.g. camphor) is weighed out into a melting-point tube that is about 70 mm long (instead of the usual 40 mm). The liquid sample is then introduced in a capillary in which it has been weighed. The end of the capillary is not sealed. The capillary consists of a tube that is about 1 mm in diameter, sealed at one end, and drawn out at the other to a fine hair-like capillary of length about 10 mm (as for determination of C and H in volatile liquids), leaving the wider portion about 9 cm in length from the sealed end. The sample is introduced into the capillary by heating and cooling as for a C and H determination but the liquid is not centrifuged to the sealed end, but instead an air bubble is left which will expel the liquid sample when the tube is heated.

For the weighing, it is contrived that the liquid remains in the hair-like capillary, but not at its extremity, so that no sample will evaporate during the weighing.

The capillary with sample is introduced, open end down, into the extra-long melting-point tube containing the solvent, which is sealed in the usual way but without heating the capillary and sample. When the sealed tube is heated in a bath of liquid all the sample will be expelled from the capillary and the determination may be completed in the usual way.

21.2 THE EBULLIOSCOPIC METHOD

In this method the molecular weight of the organic substance is determined by measuring the increase in boiling point of a known weight of solvent when a known weight of the substance is dissolved in it. The greater the molecular weight, the smaller is the increase in boiling point for equal concentrations, expressed in w/w terms.

The apparatus (Fig. 21.3) that is recommended for this purpose is that of Sucharda and Bobranski [5] which has proved, because of its design, to be particularly effective in avoiding superheating; this is, in avoiding raising the temperature of the solution above its true boiling point.

A special Beckmann thermometer is required so that the change in boiling point can be measured almost to 0.001°C (see *Apparatus* below).

To save resetting the Beckmann thermometer for the differing boiling points of the solvents used it has been found worthwhile to keep three thermometers ready set, one each for the solvents acetone, methanol and benzene.

It should be noted that some solvents such as pyridine, chloroform and water seem to be unsatisfactory for one reason or another in this apparatus, most likely because of superheating. It is therefore advisable to use only those solvents suggested here. Some workers have reported difficulty with this apparatus. The author recommends perseverance, and possibly changing the apparatus until a satisfactory one is found.

21.2.1 Apparatus

The ebullioscope. The apparatus, shown in Fig. 21.3, was designed by Sucharda and Bobranski to boil about 5 ml of solution, with superheating avoided by means of the sintered glass sealed in at G, and an undulating tube W. The boiling solution circulates and overflows at M, forming a second jacket to prevent cooling. The thermometer bulb stands in a pool of mercury in the inner cup T. The condenser jacket may be attached with rubber tubing as shown or made integral with the rest of the all-glass apparatus.

It has been found helpful to surround the apparatus with a draught screen in the form of a box provided with suitable holes for the boiler and siphon to project below. This box may be made of cardboard with the bottom sandwiched with asbestos.

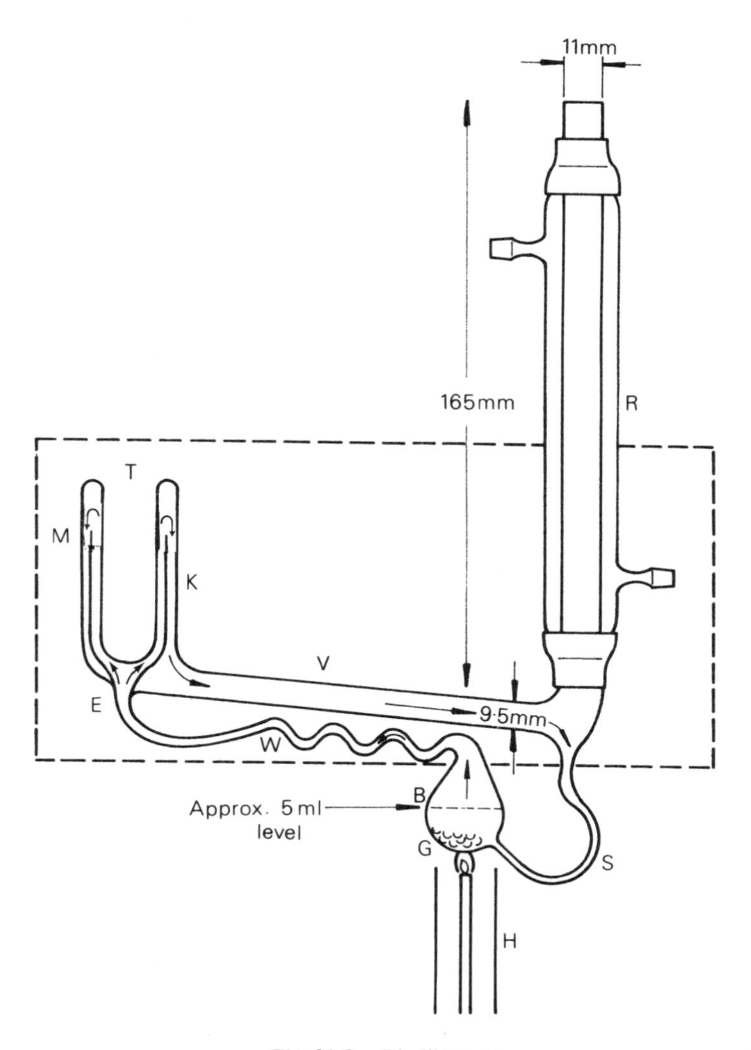

Fig. 21.3 – Ebullioscope

Beckmann thermometer. This thermometer is preferably of the small kind having a 3°C range and an overall length of about 27 cm. The more usual type (5°C or 6°C in range and about 56 cm in length) may be used provided the bulb can be immersed in mercury in the cup T.

Tablet press. A suitable press forming a tablet of $\frac{1}{8}$ in. diameter may be obtained, or a $\frac{1}{8}$ in. die from a tablet machine can be used.

Watch glasses. It saves time to have a pair of counterpoised watch glasses ready for weighing the sample.

Lens. This is needed for reading the thermometer and should be clamped at the height required.
A micro burner conforming to BS 1428 is suitable.
Device for weighing liquid samples (Fig. 21.4). This permits the liquid sample, weighed in the weighing tube (Fig. 21.5) on the hooks of the balance, to be put into the solvent in the ebullioscope without loss by contact with the condenser walls.

21.2.2 Reagents

Acetone A.R.
Methanol, anhydrous. This should be kept well-stoppered when not in use or even in an outer air-tight enclosure such as a bell-jar to prevent it picking up atmospheric moisture.
Benzene. Molecular-weight determination grade.

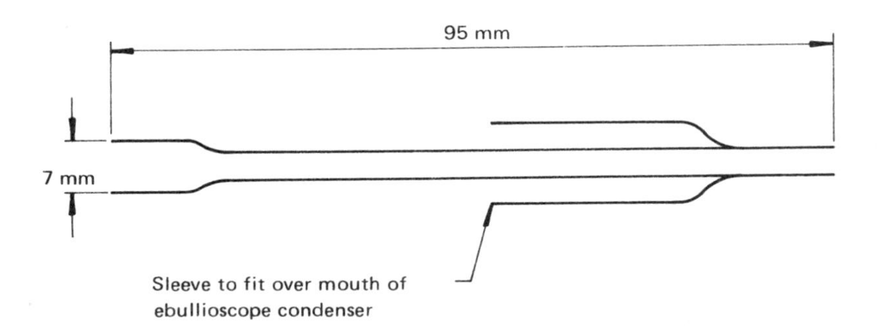

Fig. 21.4 – Glass guide for conveying weighed liquid into the ebullioscope

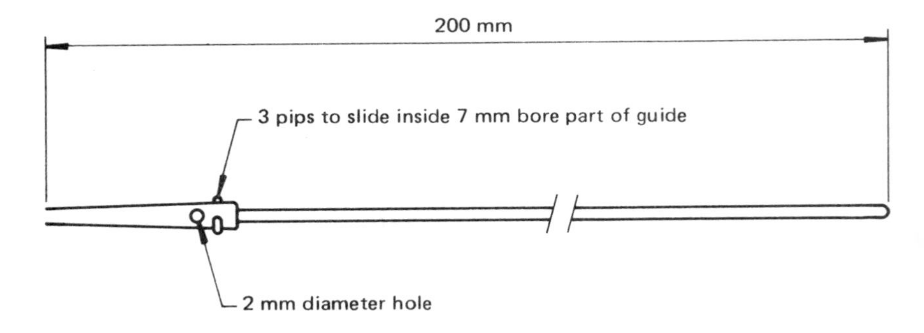

Fig. 21.5 – Tube for weighing non-volatile liquids (glass)

21.2.3 Procedure

Using a small test-tube make sure that the substance dissolves readily in the solvent, chosen in the order of preference: acetone, benzene and methanol. It will be necessary to dissolve about 50 mg in about 5 ml of boiling solvent.

Rinse the glass apparatus with acetone and finally with a little of the solvent to be used, pour out as much as possible of the solvent and dry the apparatus in an oven at 150°C and apply a rough vacuum to it to draw off all the solvent.

While the apparatus is drying prepare a pellet of about 50 mg of the sample (unless the sample is in large crystals of about 10 mg or more in which case 4 or 5 of these could be used). If a tablet press is used, use no more pressure than is needed to produce a clean tablet, so that the sample will dissolve as rapidly as possible. (Sometimes samples which appear sufficiently soluble in the preliminary test prove obstinately difficult to bring into solution in the apparatus.) Weigh the tablet (or a few crystals) of the sample on a watch glass.

Set up the dried apparatus with cold water flowing in the condenser jacket, mercury in the cup and the appropriately adjusted Beckmann thermometer immersed in the mercury cup T (Fig. 21.3). The apparatus is best enclosed in the draught-excluder box except that the boiler bulb and siphon tube should project below, as indicated in the figure. It can be supported on a metal plate clamped on the same retort stand as that supporting the glass part. This arrangement prevents direct heat from the flame from reaching the upper parts of the glass apparatus.

Now add the solvent. Although it is necessary to know the *weight* of the solvent it may be added from a 5 ml pipette. Take the temperature of the solvent at the same time so that the weight can be calculated from density tables. Bring the solvent gently to the boil and keep it boiling. This is done by applying a very small flame, from a burner equipped with a chimney, below the sintered glass at G. At first it is necessary to apply the flame intermittently, otherwise the boiler may boil dry and become overheated. Also, the solvent may circulate in the wrong direction. In this way, with care, it is possible to get the solvent to boil gently, pass up the undulating tube and overflow at the rim of the inner cup K. The liquid then runs down the sloping tube V and returns to the boiler. In effect the thermometer records the temperature of the boiling liquid and not that of its vapour. This is achieved by the quite vigorous mixing of the liquid and vapour as they ascend the undulating tube, W. In fact, when the sample is later added, a lowering of the surface tension usually results and the boiling solution tends to froth more, causing even better mixing of liquid and vapour.

When the liquid is boiling steadily the mercury thread will appear on the thermometer scale, if it is properly adjusted. If it is not quite on the scale after a few min the thermometer will require adjustment. It is unimportant at which point of the scale it finally arrives, because the subsequent rise in the reading is not likely to be more than 0.3°C.

Watch the thermometer for 5–10 min; it should give a remarkably steady reading, varying by hardly 0.001°C even if the flame is slightly increased or decreased, provided that the boiling liquid is overflowing at the inner cup (M).

Then add the weighed sample by dropping it down the condenser. Sometimes the sample will dissolve rapidly. It usually causes the boiling solution to froth more at first, as mentioned above. Then the boiling often becomes quite steady and the new boiling point can be observed after 2–3 min but the experiment should be continued for a few min longer to ensure that the value is constant.

Some solid samples do not dissolve easily, and the pellet or part of it may temporarily block the siphon tube S and the solution may circulate in the wrong direction. This may be beneficial, provided the flame is removed for a short while, because it will help to dislodge the pellet if it is in the siphon tube. However, try not to allow the solution to go off the boil, because that may favour superheating. Many substances that are difficult to dissolve may be brought into solution with patience in this way. The temperature should be observed minute by minute to confirm that all the sample is in solution and that the temperature is not still rising. When the new boiling point is steady for about 5 min the increase in boiling point, ΔT, is recorded.

For a sample that is liquid but not volatile the glass tube shown in Fig. 21.4 may be used. It is weighed empty on the pan hooks of the microchemical balance, then dipped into the liquid, when it will pick up about 50 mg of sample, and is weighed again. In the meantime the pure solvent is held at its boiling point in the usual way in the ebullioscope and the boiling point is recorded. Then, by means of the glass guide shown, the long-handled sample holder is passed down the condenser so that no sample adheres to the upper part of the condenser. After the sample has dissolved in the solvent the sample carrier may be removed and the new boiling point recorded.

Molecular weights of volatile liquids only rarely have to be determined, because the volatility of the sample is evidence that the molecular weight is low, and it can then be deduced from the results of the elemental analyses. If the determination is really necessary, weigh about 100 mg of the liquid in a weighed stoppered 10 ml standard flask, then make up to the mark with the solvent and weigh again. Determine the boiling point of the pure solvent in the usual way. Empty and dry the apparatus and determine the boiling point of 5 ml of the solution in the 10 ml standard flask. Redetermine the boiling point of the pure solvent to ensure that atmospheric pressure has not changed significantly in the time elapsed.

For precise work with any kind of sample, solid or liquid, it is recommended to use two ebullioscopes, one of which contains boiling pure solvent. If the boiling point of the pure solvent changes during the experiment this may be due to changes in atmospheric pressure and the readings from the other apparatus may be adjusted accordingly.

As a general rule, however, very precise molecular weight determinations are not required: all that is wanted is to know which simple multiple of the empirical formula weight is the correct one. For example, if 14.0% nitrogen is found then the molecular weight may be 100, 200 etc. according to the number of nitrogen atoms in the molecule. (It is surprising in practice how often 'in between' figures are found, such as, say, 150, in this example. These are probably due to impure samples.)

The molecular weight M, is calculated from the formula

$$M = \frac{K \times 1000 \times S}{L \times \Delta T}$$

where S is the weight of the sample and L is the weight of solvent in g and K is the molal constant for the solvent (1.72 deg.kg.mole^{-1} for acetone, 2.67 for benzene and 0.88 for methanol). It is best to check these figures by testing standards in each batch of solvent.

For example, suppose 37.1 mg of pure stilbene (m.w. 180) dissolved in 3.808 g of benzene gives an increase in boiling point of 0.147°C. Then

$$M = \frac{2.67 \times 1000 \times 0.0371}{3.808 \times 0.147} = 177; K = \frac{180 \times 3.808 \times 0.147}{1000 \times 0.0371} = 2.62 \text{ deg.kg.mole}^{-1}.$$

It has been found useful to prepare a graph for each solvent from which can be read a figure taking account of the constant K and the weight of the solvent at the temperature it was measured (the factor $1000K/L$).

It is possible that during the boiling some pure distilled solvent can be held at the bottom of the cool surface of the condenser. Then the effective weight of solvent is less than that actually put into the apparatus. It is such considerations which make it advisable to determine the figures for each solvent experimentally.

It should be remembered that the sample can be recovered by allowing the solvent to evaporate from the solution in an open vessel.

REFERENCES

[1] K. Rast, *Chem. Ber.*, 1922, **55**, 1051, 3727.
[2] J. Pirsch, *Chem. Ber.*, 1932, **65**, 862.
[3] J. Pirsch, *Chem. Ber.*, 1934, **67**, 1115.
[4] H. Roth, *Pregl–Roth Quantitative Organische Mikroanalyse*, Springer, Vienna, 1958, p. 328.
[5] E. Sucharda and B. Bobranski, *Semimicro-methods for the Elementary Analysis of Organic Compounds*, Gallenkamp, London, 1936.

CHAPTER 22

Boron

22.1 DISCUSSION OF THE METHOD

About 10 mg of the sample is mixed with about 20 times its weight of anhydrous sodium carbonate. The mixture is heated in platinum, gradually at first, then slowly brought to red heat so that the sodium carbonate melts and all organic material is destroyed. The melt is dissolved in a small excess of dilute sulphuric acid, boiled to remove carbon dioxide, then oxygen is bubbled through continuously to exclude carbon dioxide and the solution is titrated to a Bromocresol Purple end-point. Mannitol is then added to allow the boric acid to be titrated with 0.01*M* sodium hydroxide (which must be carbonate-free) to the same end-point. (Oxygen is used to keep out the carbon dioxide because it happens to be already available for flask combustions.)

Oxygen-flask combustion has been recommended for the determination of boron, but sometimes black specks, apparently of unburnt carbon, are left. Roth's method [1,2] of decomposition remains remarkably reliable and easy to perform for solid samples, and it is therefore preferred.

Volatile liquids should be weighed in closed capsules (e.g. of gelatine) and decomposed in a closed metal bomb with sodium peroxide, as described for silicon determination, on p. 190. The completion of the determination is then the same as for the sodium carbonate fusion described here.

This method follows very closely that described by Roth [1,2]. It is essential that the neutralization before addition of mannitol be done just as carefully as the final titration. A pH-meter can be used instead of the indicator. It should be remembered that boric acid is steam-volatile, so the acidified solution should not be boiled for longer than is necessary to remove carbon dioxide (7 sec is recommended). Carbon dioxide from the air can then be kept out by passage of oxygen; nitrogen is equally effective but oxygen is more convenient in a microanalysis laboratory because it is already available for use in oxygen-flask combustions. A quartz flask is recommended for the titration in order not to introduce boron from borosilicate glass. (The author uses 500 ml flasks for the sole reason that they are also needed for fluorine determinations, p. 108.) However, it has been found by some workers that well-used Pyrex does not contribute any boron. Boron-free glass flasks may be used, but they are not so heat-resistant as those of Pyrex or quartz.

A suitable reference substance often used is borax (stabilized in a desiccator over a saturated aqueous solution of sodium chloride and sucrose). However, boric acid that has been recrystallized and dried in air at room temperature is even more satisfactory.

Tetrafluoroborates of organic compounds (BF_4 derivatives) give difficulty if the determination of their fluorine content is attempted. The determination of boron, however, in the manner described here is easy to perform. It is true that probably monohydroxyfluoroboric acid (HBF_3OH) is being titrated and not boric acid, but the enhancement of acidity upon the addition of mannitol still takes place and permits a precise titration. In effect, if the only fluorine present is that in the BF_4 group the determination of boron is equivalent to the determination of fluorine, or at least makes the latter unnecessary.

22.2 METHOD

22.2.1 Apparatus

Platinum crucible and lid, 1 ml, with counterpoises made of Kanthal wire.
Platinum wire. Any available diameter greater than 0.5 mm, slightly flattened at one end by hammering, 1 in. long.
Silica triangle. See p. 126.
Blowpipe, for use with natural gas and compressed air.
Quartz conical flask, 500 ml with B24/29 sockets, as used for oxygen-flask combustion in the determination of fluorine. These are larger than is necessary but are available because they are used for fluorine determinations. Flasks made from boron-free glass may be used, but these must be heated more carefully to avoid cracking.
Burette, 5 ml, with polythene container for 0.01*M* sodium hydroxide. The burette is filled by squeezing the polythene container (500 ml), so only one guard-tube is needed. When not in use this is kept closed by a stopper and a pinch-clip so that the soda-asbestos does not become damp or consolidated.
Burette, 10 ml, for 0.01*M* hydrochloric acid, is an ordinary automatic type filled by pressure, so two soda-asbestos guard-tubes are needed (they are closed when not in use).
Oxygen line. Oxygen may be used if it is already available for flask combustions. However, so as not to interfere with the latter, a separate outlet and glass tube (5 mm outside diameter) are required.

22.2.2 Reagents

Sodium carbonate, anhydrous.
Mannitol.

Sodium hydroxide, pellets.

Sodium hydroxide (carbon-dioxide free), 0.01*M*. To make, first boil and cool about 750 ml of demineralized water. Put about 300 ml of this water into the polythene container of the burette. Take a pellet of sodium hydroxide in tweezers and rinse it well in a jet of water to remove any carbonate from the surface. Drop the pellet into the 500 ml container. Repeat with 2 further pellets (3 in all, i.e. 0.2–0.3 g). Dissolve and make up to about 500 ml (mark on container) with freshly boiled and cooled water. (Theoretically 0.2 g of NaOH is required.) This solution will probably have a concentration between 0.01 and 0.013*M*. To standardize it, titrate 5 ml with the 0.01*M* hydrochloric acid, with Bromocresol Purple as indicator. This solution is protected at all times with a guard-tube containing soda asbestos to prevent the entry of carbon dioxide from the air.

Hydrochloric acid 0.01*M*. Take 50 ml of 0.1*M* hydrochloric acid (available commercially) and dilute to exactly 500 ml in a standard flask with freshly boiled and cooled water. Protect with soda asbestos guard-tubes.

Sodium hydroxide, 0.1*M*.

Sulphuric acid, 2*N*.

Bromocresol Purple solution, 1%, in ethanol. Keep in a dropping bottle.

Freshly boiled and cooled demineralized water in an ordinary wash bottle.

22.2.3 Procedure

Rinse and ignite the platinum crucible, lid and wire and stand the crucible on the lid on a metal block to cool.

After 5 min for cooling, weigh the sample into the platinum crucible. Take 10 mg of sample, or up to 20 mg if the boron content is less than 3% and if enough is available. Add approximately 0.2 g of anhydrous sodium carbonate. Mix thoroughly in the crucible with the platinum wire, which is left inside. Tap down thoroughly.

Place the platinum lid on a silica triangle, stand the crucible on it and heat from below with a small micro-burner flame held close. Destroy or evaporate the organic material by heating as gently as possible, raising the temperature slowly during about 5 min. Finally just melt the sodium carbonate by heating with a small gas–air blowpipe flame and heat to redness for about 3 min. (Do not use oxygen in the blowpipe or the platinum may melt.)

Rinse a quartz flask and put 10 ml of water into it. Add to it the still warm crucible, lid and wire. Add about 2 ml of 2*N* sulphuric acid. Warm slightly and leave for 5 min for the fused mass to dissolve. Add 4 drops of Bromocresol Purple and partly neutralize with 1.0*M* sodium hydroxide added dropwise. Just complete the neutralization with the 0.1*M* alkali (to a violet colour). Add 1 drop of 2*N* sulphuric acid to make the solution acid (yellow) again.

Boil the solution over a flame for 7 sec in order to remove carbon dioxide, derived mainly from the sodium carbonate used. Immediately bubble oxygen through the solution and cool in icewater to room temperature. Pass oxygen quite rapidly and continuously until the determination is completed, to keep out CO_2. Leave the platinum crucible, lid and wire in the solution all the time.

After oxygen has bubbled for 5 min through the cold solution, nearly neutralize with 0.1*M* alkali added dropwise, and finally exactly with 0.01*M* sodium hydroxide from the burette. The solution should be neither yellow (acid) nor green or violet (alkaline), but should just show a red colour when viewed by transmitted light. One further drop of 0.01*M* alkali should produce a violet colour. Adjustment may be made, if necessary, by using the 0.01*M* hydrochloric acid from the burette.

Add 4 g of mannitol and allow it to dissolve. If the sample contains boron, boric acid will be present and the solution will become yellow again. Titrate with the 0.01*M* alkali, keeping the oxygen bubbling, until the end-point described above is just reached.

The atomic weight of boron is 10.82. Boric acid has the formula H_3BO_3 but behaves here as a monobasic acid. Hence 1 ml of 0.01*M* NaOH $\equiv$ 0.1082 mg of boron.

REFERENCES

[1] H. Roth, *Pregl-Roth Quantitative Organische Mikroanalyse,* Springer, Vienna, 1958, p. 181.

[2] H. Roth and W. Beck, *Z. Anal. Chem.*, 1954, **141**, 404, 414.

CHAPTER 23

Silicon

23.1 FUSION METHOD

The sample is mixed in a metal bomb with about 40 times its weight of sodium peroxide. The bomb is sealed and heated to 500°C. The cooled melt is dissolved in water, and after acidification, heating and filtration of the solution, quinoline silicomolybdate is precipitated by the addition of ammonium molybdate and quinoline hydrochloride solutions. The yellow precipitate is collected on tared sintered-glass crucibles and weighed after drying at 150°C.

Volatile liquids are weighed enclosed in gelatine capsules, containing about 2 mg of cellulose powder, introduced into the metal bomb containing sodium peroxide, then treated as for solids.

23.1.1 Discussion of the method

The method described follows very closely that of Christopher and Fennell [1] which is based on that of Wilson [2] except that it is on a reduced scale and that the quinoline silicomolybdate precipitated is collected and weighed instead of being titrated as described by Wilson. Christopher and Fennell found (in agreement with Armand and Berthoux [3]) that the precipitate could readily be collected, dried and weighed on sintered-glass crucibles and that the yellow solid contained the theoretical 1.200% of silicon and thus offered a very favourable conversion factor. $(C_9H_7N)_4H_4SiMo_{12}O_{40}$ (m.w. 2340.0) contains 1 atom of silicon (a.w. 28.09) i.e. 1.200% silicon.

Wilson showed that the yellow solid could be dissolved in an excess of standard sodium hydroxide solution and the surplus back-titrated with standard hydrochloric acid (in a similar way to the determination of phsophorus, p. 136). This was adapted to a reduced scale by Klimova and Vitalina [4]. It should be noted that whereas one phosphorus atom requires 26 equivalents of alkali, one atom of silicon requires only 24 equivalents because the silicic acid is not strong enough to be titrated [2].

Many very volatile liquid organic silicon compounds are known and these must be weighed in a closed vessel such as the gelatine capsule described, which must be capable of complete destruction in the metal bomb.

23.1.2 Apparatus

Metal bombs. Cups made from nickel, with a nickel lid which can be held in place by means of a strong steel nut threaded onto the main nickel part. A copper-on-asbestos ring (like a sparking-plug washer) is used to form a gas-tight closure. Such bombs were designed by Belcher and Tatlow for the determination of fluorine [5]. Alternatively the Parr bomb may be used.

Polyethylene beakers, 100 ml.

Sintered glass crucibles. A set of 3 is required for collection and weighing of quinoline silicomolybdate. One of the crucibles is used only as a counterpoise.

23.1.3 Reagents

Sodium peroxide.

Ammonium molybdate solution 5%. Dissolve 50 g of solid (A.R.) in water. Filter into a 1 litre flask, dilute to the mark and transfer to a polythene bottle.

Quinoline hydrochloride solution, 2%. Dissolve 20 ml of freshly distilled quinoline in 800 ml of hot water containing 25 ml of conc. hydrochloric acid. Allow to cool and dilute to 1 litre with pure water.

Quinoline wash-solution. Dilute 25 ml of the 2% solution to 1 litre with pure water.

Sulphuric acid, 2N. Add 27 ml of conc. sulphuric acid to 450 ml of water and dilute to 500 ml.

Carbon black suspension. Prepare some carbon black by burning benzene in a spirit lamp and collecting the carbon on a cold surface held in the flame. Make a smooth paste by mixing it with water containing a little detergent (e.g. Teepol).

23.1.4 Procedure

Thoroughly clean a metal bomb, dry it in the oven and allow it to cool. If the sample is a stable solid weigh into the bomb sufficient to contain about 0.5 mg of silicon. If the sample is a liquid or an unstable solid weigh it in a closed gelatine capsule containing about 2 mg of cellulose powder to soak up the liquid. Add 0.2 g of sodium peroxide.

Smear a thin layer of the carbon black suspension onto the faces of the bomb which are to bear on the copper washer. Insert an annealed copper washer, then close the bomb. Clamp it in a vice, then tighten the nut with a large spanner.

Heat the closed bomb in a muffle furnace for 15 min at 500°C. Allow the bomb to cool and open it. Wipe the outside of the bomb with a clean tissue and place it and the lid in a 100 ml polythene beaker. Add water gently to half-cover the bomb, cover with a polythene lid to prevent loss by spattering, and heat on a steam-bath for 15 min. Remove bomb and lid, rinsing them into the beaker with pure water. Add 10 ml of 2*N* sulphuric acid and, keeping the cover in place as much as possible, heat on the steam-bath with occasional stirring with a polythene rod, to remove carbon dioxide. Cool and filter into a 150 ml Pyrex beaker and dilute to 75 ml. Add 15 ml of 5% ammonium molybdate solution and allow to stand for 10 min. Add 5 ml of conc. hydrochloric acid and precipitate yellow quinoline silicomolybdate by adding 5 ml of 2% quinoline hydrochloride solution dropwise from a burette, stirring continuously. Heat on the steam-bath to coagulate the precipitate and cool to room temperature. Collect the precipitate on a tared sintered-glass crucible, washing 5 or 6 times with the quinoline wash-solution. Dry at 150°C for 1 hr, cool for 30 min in a desiccator and weigh against the counterpoise crucible which has been heated and cooled according to the same timetable.

The precipitate $(C_9H_7N)_4H_4SiMo_{12}O_{40}$ contains 1.200% of silicon (0.5 mg of silicon, therefore, produces about 42 mg of precipitate).

23.2 CARIUS METHOD

The sample must be solid and capable of being made into a pellet of about 20 mg, which can be weighed. In addition the sample must contain no other ash-producing substances such as metals. The pellet is sealed in a borosilicate tube with 0.26 ml of fuming nitric acid and heated at about 280°C for 3 hr. A skeleton of white silica remains in the tube with the nitric acid residues. It is collected, ignited and weighed as anhydrous SiO_2 by the procedure described below.

23.2.1 Discussion of the method

This method has been well tried and is suitable for giving an approximate silicon content if Carius equipment is available and no other ash-producing elements (e.g. metals) are present. For example, sufficiently precise results were obtained for a silicon derivative of a large molecule of an antibiotic to show that 2 silicon atoms were present in the molecule. The skeleton of silica produced from the sample is very characteristic. The method avoids the need for special equipment in situations where silicon determinations are needed only very rarely.

23.2.2 Apparatus

Carius tubes.
Tablet press.
Platinum crucible, 1 ml, and counterpoises.
Brass block.
Fine spatula, made from nichrome wire with flattened end.
Hand centrifuge.

23.2.3 Reagents

Nitric acid, fuming (A.R.).

23.2.4 Procedure

Carius tubes of length 150–200 mm and with exceptionally heavy walls (1.5 mm or more) are selected. These may be used tubes or even shorter tubes which have already been used, brought back to the required length by carefully sealing on a nearly matching piece of heavy-walled Carius tubing.

The reason for heavy-walled tubing of limited length being needed is that some solid silicon compounds are volatile and unstable and result in silica being deposited over the entire inside surface of the sealed Carius tube. As described below, it is possible to open the tube and to dry and weigh nearly all of the original tube with and without the partially hydrated SiO_2 in it. The Carius tube to be used is cleaned with water and a moist pipe-cleaner, rinsed with water and dried at 100°C.

Put into the Carius tube, after cooling, 0.26 ml of fuming nitric acid, from the safety pipette used for all Carius work. Cool the bottom 2 in. or so of the tube in solid carbon dioxide (the nitric acid solidifies) or in an ice–salt mixture.

Weigh roughly on paper about 25 mg of the sample and make a pellet in the tablet-press of about $\frac{1}{8}$ in. (3 mm) diameter. If the sample is low-melting and waxy it may be easier if the tablet press is cooled in a cold room or refrigerator before use. Weigh the tablet accurately on a flattened platinum boat, trimming it with a knife or razor blade until it weighs about 20 mg. Push the tablet carefully into the Carius tube with a flat-ended glass rod so that the pellet adheres to the tube wall without falling into the nitric acid. (The sample will then subsequently be destroyed by nitric acid vapour, and the silica will be found near the bottom of the tube.) Seal the tube firmly in the blowpipe flame so that the sealed tube has an overall length not exceeding the maximum length which allows the tube to be centrifuged. It has also usually been found that such a tube does not burst when heated and therefore the entire tube, when opened, can be weighed on a microchemical balance.

Allow the tube to stand for about 1 hr to reach room temperature. Some compounds react with the nitric acid, with rise of temperature and consequent distillation of silicon-containing portions onto the walls of the Carius tube. This is an attempt to minimize this. Finally, heat the tube for a total of 3 hr in the Carius heater with the heating current on (that is, for about $2\frac{1}{2}$ hr after the temperature reaches 280°C).

After the Carius heater has cooled to room temperature (it is usually left overnight), cool the tube in ice before releasing the pressure in the usual way by applying a flame to the tip of the capillary seal. Then cut open the tube in an oxygen–gas flame, as is done for Carius halogen determinations, to produce a test-tube with a slightly flanged mouth and about 80 mm in length. Allow the acid to evaporate by letting stand at room temperature for a few min and then by warming very gently over the steam-bath. If this is not done gently, silica may be thrown out of the tube.

Finally, dry the tube and the white deposit of silica in an oven, arranging that its temperature rises from 100 to 150°C ($\frac{1}{2}$–1 hr), then heat the tube in the Carius heater at 300°C for a further hour, raising it to this temperature in 30 min, that is, with the temperature-control at the setting used for heating sealed tubes.

It has been shown that the final weight after strong ignition will be about 90% of the weight of the white residue at this stage (i.e. it contains about 90% silica). Therefore the deposit is transferred to a weighed platinum crucible (1 ml capacity), which is strongly ignited and again weighed. However, quantitative transfer is not easy. Therefore the tube is weighed after the transfer and again after cleaning, and an estimate is obtained of how much residue was left behind. Each time the 80 mm Carius tube is weighed, both with the untransferred residue and after cleaning, it is heated and cooled according to the same timetable.

Example of Weighing

Open Carius tube (weighed against glass tube counterpoise)			*Pt crucible* (weighed against nichrome wire counterpoise)		
	mg			mg	
Opened Carius tube with residue dried at 300°C	2352.410	(A)	Ignited Pt crucible	0.116	(D)
Same, after greater part of solid transferred	2349.960	(B)	Same + solid transferred	2.440	(E)
	2.450	($C=A-B$)	Weight transferred	2.324	($F=E-D$)
Weight of sample taken	18.197	(W)			
Tube cleaned and dried	2349.540	(G)	Crucible + solid after ignition	2.200	(J)
Wt. of solid left behind	0.420	($H=B-G$)	Empty crucible	0.116	(D)
Hence total dried solid (before ignition)	2.870	($C+H$)	Wt. of ignited solid	2.084	($K=J-D$)

This would be reduced to 89.7% when ignited (see opposite) i.e. 2.87×0.897

$$= 2.574 \text{ mg SiO}_2$$

subtract blank* 0.2 mg

$$2.374 \text{ mg SiO}_2 \ (M)$$

$$\% \text{ Si in sample} = \frac{M \times 46.75}{W}$$

$$= \frac{2.374 \times 46.75}{18.197} = 6.10\% \text{ Si}$$

Fraction of ignited solid obtained from original solid

$$= \frac{100K}{F} = \frac{2.084 \times 100}{2.324} = 89.7\%$$

*Blank arrived at from several blank determinations

Clean the 1 ml platinum crucible and ignite it for 1 min in a strong Méker-burner flame, cool on a brass block for several min and weigh exactly 10 min after ignition. Cool the Carius tube and white residue (after the heating at 300°C), in a P_2O_5 desiccator for 45 min, then wipe it gently with a chamois leather and put it on the brass block for a further 15 min before weighing. Weigh it against a similar piece of tubing and metal weights. This Carius-tube counterpoise must be permanently kept in the balance-case and carefully labelled.

Transfer as much as possible of the white solid to the weighed platinum crucible by carefully loosening it with a very fine spatula made from nichrome wire, and inverting the tube with its mouth touching the mouth of the crucible and tapping (try not to allow any of the very light silica to be blown about and lost). The tube is scraped and tapped repeatedly, until no more can be transferred. Weigh the platinum crucible with the white solid to obtain its weight. Verify the weight of solid transferred by weighing the glass tube after the transfer.

Ignite and weigh the platinum crucible containing the silica according to the timetable used for the empty crucible.

Wash the glass tube with water and brush it inside at the same time with a wet pipe cleaner in order to remove all the adhering silica. Wash it well with pure water, dry it, finally at 300°C as before, and weigh the cleaned tube after the same cooling schedule. The difference will show how much silica was left behind after the transfer to the platinum crucible.

To summarize, most of the hydrated silica in the Carius tube is transferred to a platinum crucible and ignited and weighed as anhydrous silica (SiO_2). A rough estimate is then made of the amount of silica left behind (about 15% of the total), by drying the tube and weighing it and finally cleaning, drying and weighing it empty. The solid left after drying at 300°C may be shown to contain about 90% SiO_2.

REFERENCES

[1] A. J. Christopher and T. R. F. W. Fennell, *Talanta,* 1965, **12**, 1003.
[2] H. N. Wilson, *Analyst,* 1949, **74**, 243.
[3] M. Armand and J. Berthoux, *Anal. Chim. Acta,* 1953, **8**, 510.
[4] V. A. Klimova and M. D. Vitalina, *Anal. Abstr.,* 1964, **11**, 5539.
[5] R. Belcher and J. C. Tatlow, *Analyst,* 1951, **76**, 593.

CHAPTER 24

Active hydrogen

24.1 DISCUSSION OF THE METHOD

In this method [1], about 20 mg of sample is dissolved in 2 ml of anisole or pyridine in one limb of a 2-limbed flask and 2 ml of Grignard reagent (methylmagnesium iodide) is placed in the other limb. The flask is connected to a gas-measuring burette, the air is replaced by dry nitrogen and the retaining solution is set to about zero. The solutions in the two limbs are mixed and the increase in volume (up to about 8 ml) is found. The solution may be heated to 100°C and cooled and the measurement repeated.

Because water contains about 11% of active hydrogen, all solvents and apparatus must be specially dried and account must be taken of blank determinations.

The Grignard reagent, methylmagnesium iodide, will generate one molecule of methane from each atom of active hydrogen present in the molecule of a substance in solution, and, if the amount of methane generated can be measured, the active hydrogen content can be determined. Generally speaking, any hydrogen atom not attached to carbon may be regarded as active, but those attached to oxygen as in alcohols, acids and, of course, water are especially so.

The method described here is relatively simple and has been used from time to time by the author. It can often give useful information and, of course, an observation that no active hydrogen is present can also be useful.

Although an excellent solvent for this purpose, analytical-grade pyridine, even if carefully dried and distilled, may still contain active hydrogen probably in the form of methylpyridines (picolines). These remain in solution when the less soluble perchlorate is formed and the purification of pyridine in this way, as described here [2], is very satisfactory.

Pyridine (purified as described) may be used in place of anisole. The mixture should be cooled immediately after the Grignard reagent is added. Frequently a higher active hydrogen content is found in pyridine after heating than by the cold reaction only. For each solvent a blank value must be carefully determined. (Typical values are anisole, cold, nil, with heating 0.37 ml; pyridine, cold,

0.41 ml, after heating 0.90 ml.) Amyl ether may be used as the solvent for the Grignard reagent, which is then said to be more stable.

Instead of Grignard reagent, lithium aluminium hydride [3] has been used. Each atom of active hydrogen then produces one molecule of hydrogen. A similar apparatus may be used.

A very accurate method suitable for all types of sample, including polymers with very low active hydrogen content, is based on thermal conductivity measurement of the hydrogen released by reaction with aluminium diethyl dihydride [4].

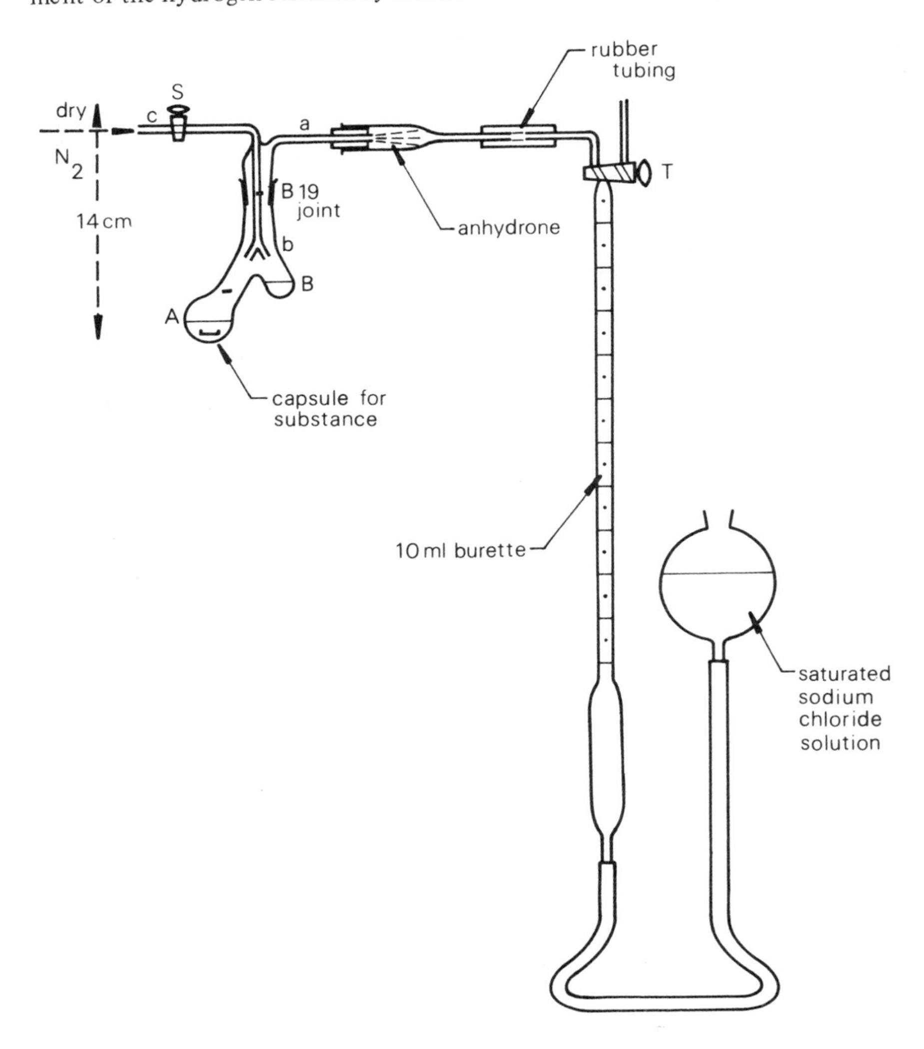

Fig. 24.1 – The determination of active hydrogen

24.2 METHOD

24.2.1 Apparatus

Gas-measuring device. This is a 10 ml burette with a two-way tap T at the top (see Fig. 24.1) and a bulb at the bottom connected by means of flexible tubing to a levelling bulb. The containing liquid is saturated sodium chloride solution (although mercury may be used).

The reaction vessel. This is a 2-limbed vessel as shown in Fig. 24.1, provided with a B19 socket. The larger limb, A, has a capacity of about 10 ml and the smaller, B, of about 5 ml. The reaction vessel is attached in use to a B19 cone by means of two strong springs on wire hooks, and the joint is greased with a little high-vacuum silicone grease. The B19 cone carries a gas-inlet tube which is forked at its lower end, to facilitate the replacement of air by nitrogen. The tube through which the nitrogen enters is provided with a glass tap S. The outlet tube is connected to the burette with rubber tubing, through an anhydrone tube, to prevent water vapour from the salt solution from reacting with the Grignard reagent in the reaction vessel.

Pipette for Grignard reagent. This is specially made with a 2 ml bulb with a second bulb above it containing anhydrone between cotton wool plugs. The pipette is filled and emptied into the reaction vessel by using a rubber-bulb pipette filler. (Because of the need for it to be absolutely dry, it is not cleaned after use, but the tip of the pipette may have to be unblocked with a wire.)

Nitrogen purification train. Nitrogen from a cylinder with a reducing valve passes through a safety valve and bubbler containing 9*M* sulphuric acid. It then passes through a large drying tube of silica gel and finally through a short tube of phosphorus pentoxide on glass wool. The nitrogen may also be passed through acidified chromic sulphate solution kept over zinc amalgam in a Drechsel bottle (i.e. through chromous sulphate) in order to remove oxygen before it enters the drying tubes.

24.2.2 Reagents

Anisole. Distil about 300 ml of anisole at ordinary pressure from a 500 ml round-bottomed flask by heating over an 'electric bunsen' in an all-glass distillation apparatus which has been thoroughly dried. The flask must contain pieces of porous pot and about 2 g of sodium metal. When the temperature is high enough to melt the sodium, shake the flask vigorously to break the sodium into many pieces. Reject the first part of the distillate. The boiling range of the main fraction, which is kept, should not be more than 1°C. Store the purified anisole in a well-stoppered flask containing a few pieces of freshly cut sodium.

Pyridine [2]. To 80 ml of pyridine add slowly, keeping the mixture cool under the tap, 340 ml (2 equivalents) of 60% perchloric acid. White pyridine perchlorate separates as a paste. Collect the white solid on a Buchner filter and press down well. Prepare some 50% sodium hydroxide solution from 40 g of sodium hydroxide

(1 equivalent) and 80 ml of water. Put the pyridine perchlorate, still damp, into a 250-ml round-bottomed flask and add the sodium hydroxide solution with shaking. Set up an apparatus for vacuum distillation, with a capillary air-leak. Distil under water-pump vacuum by heating on the steam-bath until no more distils. Add solid potassium hydroxide to the distillate, which consists of pyridine and water, until no more dissolves. Separate the upper pyridine layer, leave it to stand overnight with solid potassium hydroxide, again overnight with barium oxide, then distil it at ordinary pressure over the electric bunsen from more barium oxide. Porous pot should be added and the apparatus should be thoroughly dried. Store over a little barium oxide. About 35 ml of pure pyridine should be obtained.

Methylmagnesium iodide (Grignard reagent). Into a very dry 100 ml round-bottomed flask put 4 g of dried magnesium turnings, and connect it to a distillation apparatus tilted for reflux, and arranged so that dried nitrogen is passed over the magnesium. Fit a silica-gel drying tube to the receiver end of the apparatus.

Add 40 ml of dried anisole followed by 14 g (6.2 ml) of redistilled methyl iodide. Add a very small crystal of iodine. The start of the reaction will be indicated by the solvent becoming warm and it usually becomes so rapid that the anisole will boil under reflux. After the original reflux has subsided, immerse the flask in hot water and raise the temperature of the water (hot-plate) to 100°C and keep it there for 1 hr, with nitrogen passing slowly through. After 1 hr move the apparatus to the distillation position and pass dry nitrogen through for 2 hr at 100°C to remove the excess of methyl iodide. The reagent may be stored indefinitely in sealed glass ampoules under nitrogen.

Methyl iodide. This is redistilled and only the middle fraction is used.

Magnesium turnings. A grade described as suitable for Grignard reactions is used, after storage for at least a day over phosphorus pentoxide in a vacuum.

Barium oxide. Old stock should not be used. Water dropped on to a little should react with obvious evolution of heat.

24.2.3 Procedure

Check that dry, redistilled anisole (and pyridine if needed) and fresh Grignard reagent are available (if not, prepare them as described above). Check that the reaction vessel and weighing cup are in the drying oven and that nitrogen is flowing through its drying train (and purification bubbler, if used). Prepare a 2 litre beaker of water at room temperature and put another to boil on the hot-plate.

Attach the reaction vessel, while still hot from the drying oven, to the apparatus with two springs and a little high-vacuum silicone grease, and pass dry nitrogen through it until it is cold. The nitrogen bubbles out of the levelling bulb (in its low position). Weigh about 20 mg of the sample in a small glass cup that has been dried in the oven and cooled in a desiccator. (The sample should be thoroughly dried, unless account is to be taken of the water content of a hydrate.) Detach the reaction vessel and put the sample and cup into the longer limb. Add 2.5 ml of anisole to the sample in the longer limb, with a dried Pasteur pipette. Attach the reaction vessel again and dissolve the sample, with warming, if necessary. Remove the reaction vessel again and put 2 ml of the Grignard reagent carefully into the shorter limb from the special pipette. Reattach the reaction vessel. Allow the nitrogen to flow for a further 5 min. Turn off the nitrogen and close the glass tap (S in figure). Surround the reaction vessel up to the joint with water at room temperature. After 10 min adjust the level of the salt solution to near the zero mark with both surfaces at the same level (by raising the bulb and turning the tap (T) into suitable positions.) With the tap, T, in the position which connects the reaction vessel with the burette take the reading (near zero). By tilting the reaction vessel, add the Grignard reagent to the solution of the sample and shake well.

Re-immerse the reaction vessel in the cold water and level the salt solution in the measuring vessel and levelling tube. After occasional shaking during 10 min, read the volume of gas. Change the cold water-bath for one of boiling water. Leave, with occasional shaking, for 10 min. Re-apply the cold water-bath and take the reading again after 10 min.

The percentage of active hydrogen is given by:

$$\%\text{ active H} = \frac{V(\text{STP})}{W} \times \frac{1.008}{22.37} \times 100 \quad \text{where } V(\text{STP}) = \frac{V\text{x}273}{273+T} \times \frac{P}{760}$$

and V is the observed increase in volume, T°C is the temperature of the cold water-bath, P is the barometric pressure (mmHg) and W mg is the weight of sample taken.

Example

Benzoic acid (32.4 mg) in anisole, gave an increase in volume of 6.41 ml (after heating), at 25°C and 764 mmHg.

$$V(\text{STP}) = \frac{6.41}{32.4} \times \frac{273}{298} \times \frac{764}{760} \times 100$$

$$= 0.82_4\% \text{ of active H.}$$

$$\text{Theory: } \frac{1.008 \times 100}{122.13} = 0.825\% \text{ (m.w. of benzoic acid} = 122.13)$$

Note

The following method [5] for purifying pyridine is said to be better than the one given here. Collect pyridine distilling at 114–116°C, after drying over potassium hydroxide. Boil for 1 hr under reflex and with stirring, in 125 ml portions, with selenium dioxide. The amount required may be found by a small trial portion by weighing the selenium produced. Distil the product; the middle fraction, if dried for a week over barium oxide, gives a negligible blank.

REFERENCES

[1] T. Zerewitinoff, *Chem. Ber.*, 1907, **40**, 2023; 1908, **41**, 2233; 1909, **42**, 4802; 1910, **43**, 3590 and 1914, **47**, 1659 and 2417.

[2] F. Arndt and P. Nachtway, *Chem. Ber.*, 1926, **59**, 448.

[3] A. E. Finholt, A. C. Bond and H. J. Schlesinger, *J. Am. Chem. Soc.*, 1947, **69**, 1199.

[4] F. Ehrenberger and E. Thiemer, *Mikrochim Acta*, 1976 **I**, 33.

[5] D. Jerchel and E. Bauer, *Angew. Chem.*, 1956, **68**, 61.

Appendix A

List of text books

1 E. Sucharda and B. Bobranski, Semimicro-methods for the Elementary Analysis of Organic Compounds (Translated G. W. Ferguson), Gallenkamp, London 1936.

2 J. B. Niederl and V. Niederl, Micromethods of Quantitative Organic Elementary Analysis, Wiley, New York 1938.

3 E. P. Clark, Semimicro Quantitative Organic Analysis, Academic Press, New York 1943.

4 J. Grant, Quantitative Organic Microanalysis (Pregl), Churchill, London 1951.

5 J. Mitchell Jr., I. M.Kolthoff, E. S. Proskauer and A. Weissberger (eds.), Organic Analysis, Vols. 1–4, Interscience, New York 1953–1960.

6 R. Belcher and A. L. Godbert, Semimicro Quantitative Organic Analysis, Longmans Green, London 2nd Ed. 1954.

7 S. J. Clark, Quantitative Methods of Organic Microanalysis, Butterworths, London 1956.

8 H. Roth, Pregl–Roth Quantitative Organische Mikroanalyse, Springer, Vienna 7th Ed. 1958.

9 A. Steyermark, Quantative Organic Microanalysis, Academic Press, New York 2nd Ed. 1961.

10 G. Ingram, Methods of Organic Elemental Analysis, Chapman & Hall, London 1962.

11 N. D. Cheronis and T. S. Ma, Organic Functional Group Analysis, Interscience, New York 1964.

12 M. R. F. Ashworth, Titrimetric Organic Analysis, Vols. I and II, Interscience, New York 1964–1965.

13 R. Belcher, Submicro Methods of Organic Analysis, Elsevier, Amsterdam 1966.

14 J. P. Dixon, Modern Methods of Organic Microanalysis, Van Nostrand, London 1968.

15 R. Belcher (ed.), Instrumental Organic Elemental Analysis, Academic Press, London 1977.
16 British Standard Specification 1428.
17 T. S. Ma and R. C. Rittner, Modern Organic Elemental Analysis, Dekker, New York 1979.

Index

T

U

V

W

Z